AF764355

To

From
Bob Bull

Corning
2/2/95
(Groundhog Day!)

SCIENTIFIC RESEARCH
IN
EARLY CHINESE GLASS

SCIENTIFIC RESEARCH
IN
EARLY CHINESE GLASS

Edited by
Robert H. Brill and John H. Martin

Proceedings of The Archaeometry of Glass Sessions
of the
1984 International Symposium on Glass
Beijing, September 7, 1984
with
Supplementary Papers

A PUBLICATION OF THE CORNING MUSEUM OF GLASS

Symposium sponsored by TC17: The Archaeometry of Glass,
a Technical Committee of the International Commission on Glass

On behalf of the authors and future readers of this book, the editors express thanks to The Woodcock Foundation for funding a major portion of its publication. The publication was also supported in part by a grant from The National Science Foundation.

Published by
The Corning Museum of Glass
Corning, New York 14830-2253

Copyright © 1991
All rights reserved

ISBN no. 0-87290-126-2

Printed by
Elmira Quality Printers
Elmira, New York 14901

Cover imprint: Boli, *one of the ancient Chinese words for glass.*
Calligraphy by James C. Y. Watt.

Table of Contents

Note: Chinese names are spelled in English as specified by the authors themselves.

INTRODUCTION

Thanks to archaeology and numerous scientific investigations, much is known about ancient glass and glassmaking in the West, that is to say, in Egypt, Mesopotamia, and the Hellenistic, Roman, Islamic, and Mediaeval Worlds. We know what glass was like from its inception *ca.* 1500 B.C., how it was used, where it was traded, and even the raw materials from which it was made. In broad terms, the development of the technologies for forming glass, coloring it, and decorating it have all been traced. We have some understanding, too, of the way glassmaking evolved from each of these cultures or civilizations to the next. By comparison, relatively little is known about Chinese glass and about the role it played in the overall unfolding of glass history on a worldwide basis.

From its apparent beginnings in the Warring States Period (5th-3rd centuries B.C.), and throughout antiquity, glass was never made in China in the same great quantities as it was in the Western World. However, that early Chinese glass which has been recovered is often associated with interesting archaeological problems and raises some truly intriguing questions in the history of technology. Chinese glass is at once a study of an art and technology within itself and a reflection of contacts with civilizations outside China. Some of the most interesting questions about Chinese glass, both broad and specific, are:

1. How can one distinguish between glasses made in China and those imported from elsewhere?
2. What is the earliest glass found in China?
3. When was glass first made in China?
4. When did glass blowing first appear in China?
5. Where were the early Chinese glass factories located?
6. Was glassmaking discovered in China independently from its discovery in the Near East?
7. What circumstances led to the discovery of glass in China — or to the introduction of glassmaking from elsewhere?
8. How was glass technology in China related to other early pyrotechnologies: metallurgy, the manufacture of faience, glassy faience, glazed ceramics, etc.?
9. What chemical formulations were used for making glass in China during different historical periods?
10. How do chemical compositions of Chinese glasses compare to those of "Western" glasses?
11. What materials were used for making glass in China?
12. What accounts for the presence of barium in the earliest Chinese glasses?
13. What chemical formulations were used in different parts of China and during different historical periods? and what can be inferred from any differences in those formulations?
14. How are the 11th-10th century B.C. faience beads excavated in western China related to early Chinese glass?
15. How does the technology of Chinese blue and Chinese purple relate to Chinese glass technology and what implications does that have for connections with Middle Eastern pyrotechnology?
16. What role did Chinese glass play in the overall unfolding of glass history worldwide?

A few pioneering studies carried out in the 1930's established that there was something different and something special about the glass of ancient China. In the years which followed, the field lay more-or-less dormant, although in the fifties and sixties renewed archaeological excavations began to produce additional glass finds. The finds were few in number it is true, but nevertheless they stimulated the interest of Chinese archaeologists who recognized that not only did glass provide valuable clues for dating, but it also offered important evidence for studying a heretofore overlooked aspect of Chinese art and technology. In 1982, one of the editors (R.H.B.) was able to locate only three Chinese scholars who had more than a passing interest in early Chinese glass. By the time of the 1984 International Symposium on Glass in Beijing, the papers of which are assembled here, some fifteen individuals or groups had become engaged in serious glass studies. Interestingly, the study of Chinese glasses among Chinese scholars has been promoted largely by chemists and other physical scientists who wished to learn more about China's heritage in glass. In other countries, studies by art historians and archaeologists have led the way.

As a result of these studies, answers to some of the questions listed above have begun to emerge — and some new questions have appeared.

This book is based largely on the *Archaeometry of Glass* sessions of the 1984 International Symposium on Glass, held in connection with the Annual Meeting of the International Commission on Glass in Beijing. The symposium was presented under the auspices of the Chinese Silicate Society. It was supported by The Chinese Association for Science and Technology, The China State Administration of Building Materials Industry, The Academia Sinica, and The China Ministry of Light Industry. The *Archaeometry of Glass* sessions (Session AE) were held on Friday September 7, 1984. The sessions also marked the first formal meeting of TC17: The Archaeometry of Glass, a technical committee of the International Commission on Glass.

Although only seven papers were scheduled on the printed program, eleven papers were presented. Two of the original papers were read in title only owing to the unfortunate absence of two authors who were unable to attend the sessions. We are pleased to be able to include both of these papers in the present volume. In all, about sixty persons attended the sessions. There were lengthy, valuable discussions and cordial, open exchanges of information, greatly facilitated by the simultaneous English/Chinese translations of Mr. Liu Tongzhau.

In every respect, the meeting was a resounding success, exceeding by far everyone's highest expectations. It proved so successful, in fact, that a previous commitment to publish the proceedings along with the rest of the Symposium papers was withdrawn by the publisher shortly after the meeting because there were more papers presented than they had space for. Consequently, the meeting's only shortcoming — its delay in publication — resulted from its having been too successful!

Early in 1986, Prof. Gan Fuxi of the Shanghai Institute of Optics and Fine Mechanics, Academia Sinica, a Co-chairman of the Organizing Committee of the Symposium, published several of the Archaeometry of Glass papers in Chinese.

Because this publication was delayed, it seemed wise to turn the delay to the readers' advantage by adding some supplementary papers to those included as actual parts of the 1984 Symposium. The contents of the book are therefore divided into two parts: the Symposium Papers (including the two read only by title) and Supplementary Papers.

Either the individual authors or Prof. Gan submitted typescripts of English translations of the papers originally written in Chinese. Maxine G. Smith and Ruth Sprague made a major contribution by retyping all of the papers onto a computer disk and coping with the myriad difficulties of arranging tables and texts into a consistent format. Richard W. Price, also of The Corning Museum of Glass, assisted in the initial editing stages and James C. Y. Watt, of the Metropolitan Museum of Art, provided continuing advice throughout the entire publication process.

Primarily, the editors have modified these typescripts only where necessary in order to improve English spelling, grammar, general reada-

bility, and clarity. Every effort has been made to retain the authors' intentions as accurately as possible within the editors' own understanding of the typescripts. Some inaccuracies in spellings of Chinese words must inevitably have been inadvertently introduced but we hope that will not affect the readers' understanding. In particular, references to early Chinese writings have been entered just as the authors submitted them, for to have attempted to systematize them would have been far beyond the editors' capabilities.

Where the editors were uncertain, question marks or brief notes have been inserted in square brackets {}. Having had the advantage of seeing the entire publication in overview, the editors have taken the liberty of inserting occasional Editor's Notes, also enclosed in square brackets {}. These have been added only in instances where we felt they would be to the readers' benefit. There is a certain amount of redundancy in the papers, but it was decided not to cut too deeply for fear of interfering with the coherence and internal logic of the individual contributions. Finally, there are a few instances where the editors are in disagreement with opinions expressed or interpretations of findings, but, as is only proper, these have been allowed to stand just as expressed by the authors and without editorial comment.

It seems fair, and we hope not immodest, to say that the knowledge of Chinese glass assembled by the authors of this book, and the new insight they have provided, have doubled our understanding of the subject. But in addition, the research reported here is of very special importance to Western scholars who have an interest either in glass *per se* or in things Chinese. Until the 1984 Symposium, the literature on Chinese glass available in Western languages consisted almost exclusively of the thinking of Western scholars. Only now do glass historians and historians of technology outside China have an opportunity to know just how the Chinese themselves view their own glass history.

The main value of this Symposium may lie, ultimately, not so much in the findings presented, as it will in the excitement generated among the participants. Our findings, after all, are subject to change, and our still sketchy picture of Chinese glass is bound to improve as excavations and investigations continue. Who knows what marvelous new discoveries and unexpected new insights lie ahead?

Robert H. Brill and John H. Martin
Corning, New York
December, 1990

PART I

SYMPOSIUM PAPERS

INTRODUCTION TO THE SYMPOSIUM PAPERS

GAN FUXI

Shanghai Institute of Optics and Fine Mechanics, Shanghai

Research on ancient Chinese glasses, prompted by recent archaeological discoveries, is attracting growing attention on the part of scholars both in China and abroad. The papers presented at the *Archaeometry of Glass* sessions of the 1984 International Symposium on Glass reflect the rapid progress that is being made. At this first international meeting ever held on Chinese glass, experts from China and other countries presented their findings to one another and held extensive discussions on the origin and development of Chinese glass. A great deal of new information was exchanged, and the success of the symposium was the result of the devotion of the researchers and of the applications of modern analytical techniques. With continued cooperation of archaeologists and glass scientists, a new era for glass studies is underway.

Many glass experts have started their research by looking at early Chinese writings. Among the words for glass in these writings are *miaolinlangan*, *liuli*, *liulin*, and *boli*. According to Zhang Hongzhao, *liuli* was derived from the pronunciation of *liulin*[1], and a new suggestion for the derivation of *boli* has recently appeared.[2] Some of the historical Chinese books which contain information about glass are: *Lunheng*; *Yugong* in *Shangshu*; *Xiyu zhuan* and *Dili zhi* in *Han shu*; *Yantie lun*; *Xiyu* in *Weishu*: *Taiping yulan*; *Jiu Tangshu*: and*Dayuezhi zhuan* in *Baishi*.

Glass scholars have always been interested in connections between Chinese glass and that of other countries. For example, Benrath[3] reported that a Middle Kingdom tomb in Egypt contained glass bottles bearing Chinese characters. But most researchers were preoccupied with glass that moved eastward into China. Neumann[4] believed that glass was imported into China from Egypt, Rome, and the Middle East. Farmakovski[5] stated that there was no glassmaking in ancient China but that frit preparation probably existed in connection with ceramic glazing. In the 1930s, however, Western researchers realized that glass was being made in China during the Han and Tang Dynasties, although glass blowing did not begin in China until much later.[6] The findings of the participants in our Symposium have verified that glass was, indeed, made in China as early as the Warring States Period (the 5th century B.C.) and that some of the glass found in China was imported.

When chemists began analyzing ancient glasses, a few Chinese examples were included. Among the pioneers in this field were W.E.S. Turner in England,[7] M.A. Besborodov in Russia,[8] and Kazuo Yamasaki in Japan. Prof. Yamasaki participated in this Symposium.

Many of the speakers at the Symposium referred to the work of Beck, Seligman, and Ritchie, who, in 1934, were the first to note that the earliest Chinese glasses contained high lead and barium contents.[9,10]. Joseph Needham,[11] the eminent historian of Chinese science, noted — as did Turner, Besborodov, Yamasaki, Brill, and other authors who considered the topic — that barium was not used for glassmaking in the West until many centuries later.

T. Maruyama,[12] Watanabe,[13] and Yamasaki have all made thorough studies of East Asian glassmaking and established relationships between Chinese, Japanese, and Korean glasses. Yamasaki believed that glass exchange between China and Japan began in the Han Dynasty. In Korea,[14] glasses dated ca. 200 B.C. have been found, and they are believed to have come from China. Brill[15] and Yamasaki[16] have recently analyzed several kinds of Chinese glasses, adding a large number of lead and lead-barium glasses to the previously known examples of these types of glasses. This was the state of knowledge about Chinese glass as it existed outside China until the 1984 Symposium.

Within China itself, inquiries into the origin of Chinese glass began in the 1930s. Attention was directed mainly toward analyses of historical documents.[17,18,19] In the 1950s, Yuan Hanqing carried out valuable chemical analyses of Chinese glasses,[20] and he also concluded that the earliest Chinese glasses were characterized by the presence of both lead and barium. Yuan worked mainly with excavated glasses of the Warring States Period. Excavations during the past thirty years have provided a growing number of glasses for research and led to the investigations reported at the 1984 Symposium.

In the 1970s, the author arrived at the "self-invention" hypothesis, namely, that Chinese glassmaking had originated in China.[21] This touched off a dispute among researchers in the field. Chemical analyses of excavated relics done for Yang Boda[22] supported the hypothesis. However, Zhang Fukang *et al.*[23] also analyzed some opaque beads and tubes, and they learned that those samples were so high in crystalline quartz that they should be considered faience instead of glass.

We have made a careful review of the analyses published by several authors throughout this book[24] and combined them with the analyses of Yuan Hanqing. It appears to us, as it has to the authors themselves, that the development of Chinese glass can be traced according to chemical compositions. In our interpretation, we see four separate stages.

1. From the Western Zhou Dynasty (ca. 1100 - 771 B.C.) to the Spring and Autumn Period (770 - 476 B.C.): Glasses of this period are mainly beads and sword ornaments. The chemical compositions of the beads are dominated by SiO_2, and there are large amounts of crystalline quartz present. Strictly speaking, this material should not be called glass, but instead faience, as found in other parts of the world.[25] But the sword ornaments are transparent glasses, generally calcium-silicate glass ($X\text{-}CaO\text{-}SiO_2$) or potassium-calcium-silicate glass ($K_2O\text{-}CaO\text{-}SiO_2$).[26]

2. From the Warring States Period (475-227 B.C.) to the Sui Dynasty (581-618 A.D.): Beads, ear spools, sword ornaments, cups, and bowls of this period have been unearthed. In this time, distinctive glass systems were adopted: early in the period, the $PbO\text{-}BaO\text{-}SiO_2$ system dominated, but later, it was the $PbO\text{-}SiO_2$ system. Occasionally, $K_2O\text{-}SiO_2$ glasses appeared.

3. From the Tang Dynasty (618-907 A.D.) to the Yuan Dynasty (1271-1368 A.D.): The glasses were mainly utensils. Compositions included $PbO\text{-}SiO_2$, $K_2O\text{-}CaO\text{-}SiO_2$, $Na_2O\text{-}CaO\text{-}SiO_2$, and $Na_2O\text{-}CaO\text{-}PbO\text{-}SiO_2$.

4. From the Ming Dynasty (1368-1644 A.D.) to the Qing Dynasty (1644-1911 A.D.): There are many types of glass containers such as bottles, cups, and jars. The main compositions of the glasses were $K_2O\text{-}PbO\text{-}SiO_2$, $K_2O\text{-}CaO\text{-}SiO_2$, $Na_2O\text{-}CaO\text{-}SiO_2$, and $Na_2O\text{-}CaO\text{-}PbO\text{-}SiO_2$.

Concerning the earliest glasses, barium may have been added in order to imitate jade.[27] Another reason may have been that lead and barium minerals occur together in Henan and Hunan, where most glasses were produced at that time. In those areas where lead ores were scarce (for example, Guangdong and Guangxi), the glasses were $K_2O\text{-}SiO_2$ types (containing 15% K_2O, only rarely

Na_2O and CaO). This latter glass system is believed to be unique to China because it has not yet been found in other countries.

It is well known that introducing PbO and K_2O into a glass can effectively lower its melting temperature, inhibit devitrification, and produce desirable viscosity characteristics. This, apparently, was known by glassmakers in ancient China. It is also scientifically reasonable that due to differences in the raw materials available in different areas, glass compositions could have varied from place to place.

In closing, we express our hope that the findings reported at the 1984 Symposium will stimulate further inquiry into the origin and evolution of glassmaking in China. A great amount of work remains to be done. We thank Dr. R. H. Brill and Dr. J. H. Martin (ret.), both of The Corning Museum of Glass, for their help in editing the English edition of the proceedings, and Mrs. Lin Fengyin, who did most of the translations. Without their help, the proceedings would never have been published.

Notes

1. Zhang Hongzhao, *Shiya*, Chinese Institute of Geology, 1921.
2. Robert H. Brill, "Some Thoughts on the Origin of the Chinese Word 'Boli'," *Glass and Enamel*, 1990, v. 18, no. 5. (In Chinese.)
3. H. Benrath, *Die Glasfabrikation*, Braunschweig, 1880, p. 2.
4. B. Neumann, "Antike Glaser, II," *Zeitschrift für Angewandte Chemie*, v. 40, no. 40, 1927, p. 363.
5. M. V. Farmakovski, *Rimskie steklovarennye pechi, Izvestia Instituta Arkheologicheskoi Tekhnologii*, St. Petersburg, 1922, pp. 87-88.
6. A. Salmony, "Chinese Glass, Han and Tang Periods," *Journal of the American Ceramic Society*, 1930, v. 13, no. 5, Part 1, Abstracts, p. 330.
7. W.E.S. Turner, "Studies in Ancient Glasses and Glassmaking Processes," *Journal of the Society of Glass Technology*, 1954, v. 38, p. 445T; 1956, v. 40, 162.
8. M. A. Besborodov, *Steklodelie v Drevnei Rusi, Minsk, 1956.*
9. *H. C. Beck and C. G. Seligman, "Barium in Ancient Glass," Nature*, 1934, v. 133, no. 6, p. 982.
10. C. G. Seligman, P. D. Ritchie, and H. C. Beck, "Early Chinese Glass from Pre-Han to Tang Times," *Nature*, 1936, v. 138, p. 721. {For a review of the history of glass analysis see: E. R. Caley, *Analyses of Ancient Glasses, 1790-1957*. The Corning Museum of Glass, 1962.}
11. J. Needham, *Science and Civilisation in China*, v. 4, Part I: Physics, pp. 101-105, Cambridge University Press, 1962.
12. T. Maruyama, "A Study of Ancient Glasses," *Ceramics*, 1968, v. 3; 1969, v. 4.
13. Watanabe, *Ceramics*, 1969, v. 4, no. 6, pp. 501.
14. *Korea Acta*, 1961, no. 18, p. 109.
15. R. H. Brill, K. Yamasaki, I. L. Barnes, *et al.*, "Lead Isotopes in Some Japanese and Chinese Glasses," *Ars Orientalis*, 1979, v. 11, pp. 87-109.
16. K. Yamasaki, "Chemical Compositions of Ancient Glass Found in Japan," *Proceedings of the 10th International Congress on Glass*, Kyoto, 1974, pp. 15-20.
17. Liu Dunzhen, *Journal of the Chinese Society of Architecture*, 1932, v. 3, no. 3.
18. Hu Zhaochun, *Quarterly of the Zhongshan Culture and Education Museum*, 1935, v. 2, no. 4.
19. Sen Han, *Journal of the Hebei Institute of Industry*, 1937, no. 3.
20. Yuan Hanqing, *Digest of the 1957 Annual Meeting of the Chinese Chemical Society*, pp. 80-81.
21. Gan Fuxi, *et al., Journal of the Chinese Silicate Society*, 1978, v. 6, nos. 1 and 2.
22. Yang Boda, *Cultural Relics*, 1979, no. 5.
23. Zhang Fukang, *et al., Journal of the Chinese Silicate Society*, 1983, v. 11, no. 1.
24. This volume, Chapters 1, 3, 4, 7, 10, 11, 15, and 18; also Notes 19 and 20.
25. Cheng Zhuhai, *Proceedings of the 1984 Changsha Conference on Ancient Glasses*, May 1984; and R. H. Brill, "Ancient Glass," *Scientific American*, Nov. 1963, v. 209, pp. 120-131.
26. This volume, Chapter 10.
27. This volume, Chapters 4 and 10.

CHAPTER 1

THE EARLY GLASS OF CHINA

AN JIAYAO

The Institute of Archeology,
Chinese Academy of Social Sciences,
Beijing

Scholars both in China and abroad have begun to pay great attention to the origins and evolution of glass in China. Recent archaeological excavations in China have unearthed numerous glass objects, and these have contributed much to our knowledge of the history of Chinese glass.

Many of the articles published in the past few years have dealt with the question of the origin of Chinese glass, concentrating mostly on pre-Han objects such as beads, *bi* disks, sword ornaments, and so on. Objects later than the Western Han Period, however, have been largely ignored since they cannot directly provide evidence for the origins of glass. Although these vessels and containers came into being much later than glass beads and *bi* disks, they are indicative of the development of glassmaking to a more advanced stage, and they merit further attention and research.

The earliest known Chinese glass vessels are from the Western Han Period. Glass vessels from subsequent dynasties have also been discovered, but they are few in number and the finds came from widely separated sites. Accordingly, research on them has been somewhat difficult. In order to compile enough material for this study, approximately 200 glass vessels excavated from fifty sites dating from the Western Han through the Northern Song Dynasties (206 B.C.-1127 A.D.) were collected and analyzed. Although this period encompasses more than a dozen centuries, these objects can still be considered "early glass," especially when compared with the glass of the Qing Dynasty (1644-1911 A.D.).

With the exception of twenty objects from a Xinjiang collection, the 180 vessels and fragments used in this study were all found in archaeological excavations, and many come from dated contexts. Because the composition of a glass may be useful for identifying its origin, quantitative chemical analyses of twenty-one specimens were made by the Institute of Archaeology of the Chinese Academy of Social Sciences and the Research Institute of Building Materials. The samples were collected with the help of many museums, and the glasses studied are listed in Table 1. (Of these, the author has examined 130 pieces personally.) The results of the quantitative analyses are reported in Table 2. X-ray fluorescence analyses were carried out on 31 intact specimens by Qinghua University (See Table 3), and the results of these tests provided valuable information for the chemical classification of early glass vessels of China.

Glass vessels dating from the Western Han (206 B.C.-24 A.D.) through the Northern Song (960-1127) Dynasties have always been separated according to their places of origin: domestic glass and imported glass. Our first aim was to distinguish between these on the basis of chemical analyses. Careful comparisons have thus been made between early glass vessels excavated in China and their counterparts manufactured in Western glassmaking centers. These studies have been based on type, decoration, technique, and composition. It has been found that some glass vessels are rare and unlike obvious Chinese types; instead they resemble vessels made outside China. Other vessels have unusual styles which have not been found outside China.

Imported Glass Vessels. During the interval from the Western Han to the Northern Song Dynasties, numerous glassmaking centers were located in the Mediterranean area and on the Iranian plateau. Roman, Sasanian, and Islamic glass were among the objects which came into China. Roman glass usually refers to glass manufactured in the Roman Empire from the lst century B.C. to the 5th century A.D. Glass from the eastern Mediterranean, manufactured between the fall of Rome and the rise of the Caliphate, can be considered similar to Roman glass.

During the lst century B.C., the two principal centers of glass manufacture in the Hellenistic world fell under Roman control: first the Syro-Palestinian coastline in 63 B.C., then Egypt and Alexandria in 30 B.C. It was during this period that the most significant innovation in the history of glassmaking was made: the discovery of glassblowing. This technological advance revolutionized the glass industry. The introduction of blowing had the concomitant effects of simplifying production and reducing costs, thereby making glassware available to a general market.

Some typical Roman glasses have been excavated from the tombs of the Han-Northern Song era. Fragments of a ribbed bowl (Fig. 1) unearthed from an Eastern Han tomb at Ganjiang, Jiangsu Province, exhibit a marble pattern of purple with streaks of opaque white while the exterior of the vessel has radiating, protruding ridges. The results of analysis show that it was cast from soda-lime-silica glass. This type of ribbed bowl was rare in China, but it appeared in great numbers in Italy and the Mediterranean area during the lst century A.D.. Roman glassmakers are known to have been adept at making marbled ribbed bowls, and many intact examples of this type of bowl, as well as hundreds of fragments, have been found in the West.

An intact glass cup (Fig. 2), excavated at an Eastern Jin tomb (317-420) in Xiangshan, Nanjing, Jiangsu Province, is colorless with a yellow-green tinge and a white weathering crust. A rosette with seven petals appears on the underside, and two grooves decorate the shoulder. The decoration was cut. Fragments of this kind of cut glass have been excavated at several places in the Eastern Jin tombs of Nanjing, including fragments found in the northern suburb of Nanjing and now at Nanjing University. The latter glass fragments have been analyzed, and all of them are soda-lime glasses with low iron contents; manganese dioxide was used as a decolorizer. The chemical composition of these Nanjing fragments is very close to typical Roman glass.

The Roman craftsmen were experts in the rosette design and cutting technique. Some Roman glass cups of the 3rd and 4th centuries from the eastern Mediterranean are similiar in form and decoration to the Xiangshan cup. Other vessels excavated in China, which probably originated in the Roman World, include the duck-shaped vessel, bowls, and cup from the Feng Sufu tomb at Beipiao, Lianoning Province, and the bowls from the Feng clan tombs in Jingxian, Hebei Province. The bowls from the Western Han tombs in Guangdong Province and from the Eastern Han tombs at Guangxi Province may also be Roman.

Sasanian glass was produced on the Iranian plateau between about the 3rd and 7th centuries. Its development was influenced by Roman glass, but it nonetheless achieved its own distinctive style. Some typical Sasanian glass vessels have been unearthed in China. Fragments of a cut glass bowl excavated at a Western Jin

Dynasty (265-316) tomb at Wulidun, Echeng, Hubei Province, show Sasanian characteristics. The glass is colorless and transparent, with a yellow-green tinge. Based on these fragments, one can imagine the original shape of the bowl: a round base, outsplayed rim with cut-flat edge, and protruding sides. The body of the bowl is decorated with three rows of cut facets. At the bottom there is a large facet, and at the shoulder there are two wheel-cut grooves. This is the only example of this kind of glass bowl excavated in China, but many similar bowls were unearthed from Sasanian tombs at Gilan, Iran. Another bowl of this type was excavated from Niizawa Senzuka Tumulus No. 126 at Kashihara City, Nara, Japan. It was carefully studied by Japanese scholars and is regarded as having been made in Iran.

An important Sasanian glass bowl was recently found in China (Fig. 3) in a Northern Zhou Dynasty (557-581) tomb at Guyuan in Ningxia Province. The body of the hemispherical bowl is covered with two rows of convex roundels, each roundel having a central, countersunk disc. At the bottom there is a larger, convex roundel. In addition, a green bottle (Fig. 4) with a round body and narrow neck unearthed from a Tang Dynasty tomb at Guanlin, Luoyang, and a glass bowl from the Tang Dynasty hoards at Hejiacun, Xian, are also probably Sasanian.

In the 7th century, Islam spread over the eastern Mediterranean and the Iranian plateau, thus occupying two major glass centers. The technique and form of Islamic glass continued directly along the preceding Roman and Sasanian lines, but influences brought from the far corners of the Caliphate enabled the glass to develop its own style by the 9th century.

Several vessels representative of Islamic glass have been excavated in China. Three engraved glass bottles were found in excavations at Northern Song stupa sites: stupa No. 5 at Dingxia, Hebei; the Wuwei stupa, Anhui; and the Huiguang stupa in Ruian, Zhejiang (Figs. 5 - 7). All of these bottles have relatively long necks, flat bottoms, and engraved patterns of linear and facet cutting very similar to glass forms common in the Caliphate in the 10th century. Vessels of this type, dating from the 7th to 11th centuries, have been found in excavations of Islamic sites such as Susa and Nishapur.

A transparent blue bottle (Fig. 8) unearthed at the Dingxian stupa appears to have been free blown. Some similar vessels have been found at the 10th century site of Nishapur. Finally, a green cup (Fig. 9) excavated from the Liao Dynasty (916-1125) tomb of Geng Yanyi at Dayingzi, Chaoyang, Liaoning Province, bears a handle with a thumb-ring. Islamic vessels of this type were made of metal and pottery as well as of glass.

Domestic Glass Vessels. Small and exquisite glass objects have often been found in the tombs of the Spring and Autumn Period (770-476 B.C.) and the Warring States Period (475-221 B.C.). The earliest of these objects are two double-handled cups and a plate (Figs. 10, 11) excavated from the tomb of Liu Sheng, Prince Jing of Zhongshan in the Western Han Dynasty. The double-handled cups are a typical Chinese form, and vessels of this kind were usually made of lacquer and pottery during the Western Han Dynasty. The technique and composition of these glass vessels followed the established tradition of the Warring States Period, that is, they were made of lead-barium glass.

Three gourd-shaped bottles, two other bottles, and a monk's alms-bowl (Fig. 12) found in a Northern Wei (386-534) stupa at Jingxian, Hebei Province, are the most important glass vessels found from that period. Based on the quality of the glass and the production technique, these vessels appear to have had the same origin. All of them were blown. The bottles have a round lip formed by an inward rolling of the mouth and trailed-on feet. These techniques, passed down from the Roman and Sasanian traditions, were first used in China during the Northern Wei Period. Although these pieces are representative of a relatively advanced technique, their simple, irregular shapes and the quality of the glass materials compare poorly with Near Eastern glass of that time. The forms of these vessels are traditional Chinese forms, and pottery of similar shapes can be found in China.

The Sui Dynasty lasted only 37 years (581-618), but it was an important period for glassmaking, and many fine glass vessels have been discovered. The most noteworthy are the eight pieces excavated from the tomb of Li Jingxun in Xian. These vessels may be divided into two groups. A box, two egg-shaped vessels, and a tube (Figs. 13, 14) belong to one group. All are grass-green, transparent, and formed by blowing. No pontil mark appears on the bottom. The glass box, on which x-ray fluorescence analysis has been performed, is of high lead glass without barium. The other group includes a green flat-sided bottle, a dark green vase, a pale green cup, and a sky blue cup (Figs. 15, 16). All are blown. Each has a pontil mark on the bottom. The two cups and the vase, on which x-ray fluorescence analysis has been performed, are ordinary soda-lime glass without lead or barium. The glass vessels of both groups are typical Chinese shapes, and similar porcelain vessels are often found in Sui tombs. The glass objects from the Li Jingxun tomb were probably made in China, indicating that the making of soda-lime glass vessels and use of the pontil began in China as early as the Sui Dynasty.

Of the Tang Dynasty (618-907) glass vessels, the most outstanding are the two yellow vases, a green cup, and a green bottle (Fig. 17) unearthed from the tomb of Li Tai in Yunxian, Hubei Province. The form of the yellow vases is similar to that of the pottery of the Sui and early Tang Periods. The vessels were blown, and the mouth was fire-polished to form the lip. Chemical analysis of the fragments showed that the glass contains a high percentage of lead (64% PbO). The green bottle and cups are of a shape commonly seen in Tang vessels. Chemical analysis revealed that the vessels are a soda-lime glass with higher magnesium and potassium content. Some of the most distinctive Tang Dynasty glass vessels are the small bottles for the ashes of the Buddha or Buddhist saints found in the stupas of temples. Some of these vessels, unearthed before 1949, were taken abroad. However, three vessels of this type (Fig. 18) were recently found at sites in Jinchuan, Gansu Province, and Xian and Ningan, Heilongjiang Province. All are eggshell-thin and round, with a long neck and concave bottom.

Among the glass vessels from the Northern Song Dynasty, the most numerous are gourd-shaped bottles found in the stupa foundations of temples. They include ten pieces from stupa No. 5 at Dingxian, Hebei Province; thirty-three pieces from stupa No. 6 at Dingxian; three pieces from Mixian, Henan Province; and three pieces (Fig. 19) from Lingtai, Gansu Province. X-ray fluorescence analysis on twenty pieces of gourd-shaped bottles from Dingxian revealed that all are made of lead glass. Quantitative chemical analysis has been carried out on the fragments of the gourd-shaped bottle from Lingtai. The intriguing result is that the glass contains 50.31% PbO and 10.09% K_2O, making it a potash-lead-lime-silica glass. Other interesting pieces are the glass bowls (Fig. 20) discovered in the Dingxian stupas. They are large objects with thin walls and pontil marks.

Characteristics of Domestic Chinese Glass Vessels. According to archaeological findings up to now, although glassmaking failed to become a main handicraft in ancient China, it had a long history of slow but steady development. The Northern Wei Dynasty (386-534), during which glass was blown, and the Sui Dynasty, in which soda-lime glass vessels and the pontil technique came into use, are two important periods in the history of Chinese glassmaking. This may be directly related to the arrival of glassmakers from the Central Asian trade routes, recorded in the annals of the Northern Wei Dynasty.

Most Chinese glass vessels were made of lead glass, but the actual composition of the lead glass varied in different periods. During the Han Dynasty, lead-barium glass was prevalent. The Sui and Tang Dynasties produced high lead glass, and potash-lead glass was the most common form during the Northern Song Dynasty. The earliest historical record mentioning lead glass is from the Northern Song poet Su Dongpo, who wrote, "It is deceitful to make jade by boiling lead with white stone." This passage surely refers to the lead glass vessels of that period.

Although earlier records contain direct references to lead glass, one passage, "Our glass has a more beautiful luster than that of the West," seems to point to the existence of lead glass. Lead-containing glasses have a higher refractive index than soda-lime glass.

In the early Chinese records, there are many descriptions of the use of glass. These include containers for wine, food, drugs, and relics of the Buddha. However, none of the descriptions is detailed enough to help distinguish between the uses of domestic and imported vessels. Judging from the materials available to the author, the majority of imported glass vessels were used for wine, food, and perfume, whereas most of the native glasses were artworks for enjoyment or used for Buddhist relics.

Import Routes of Foreign Glass Vessels. Like Roman gold coins and Sasanian silver coins, the foreign glass vessels unearthed in China are very important in studying the history of trade between the East and the West. The places where these vessels were discovered have provided evidence of such trade routes.

During the Han Dynasty, imported glass vessels were distributed mainly along the coast in southeastern China (for example, in Guangdong, Guangxi, and Jiangsu Provinces). Most of these vessels were Roman, which agrees with evidence found in ancient writings. The annals of the Han Dynasty record that Emperor Wu sent people across the seas to buy glass. The "Periplus of the Erythraean Sea," written probably late in the 1st century A.D., mentions Chinese products as well as the transportation of glass. It is noteworthy that in Arikamedu, the Indo-Roman trading station on the east coast of India, some fragments of typical Roman glass vessels have been found. One fragment of a ribbed bowl is similar to the fragments from the Eastern Han Dynasty tomb at Ganjiang. Another fragment of a cast bowl with grooves, found in the same area of India, is just like the bowl unearthed from an Eastern Han tomb in Guangxi Province. These archaeological findings suggest that Roman glass vessels were imported by sea.

During the Southern and Northern Dynasties (420-589), China was divided; nonetheless glass vessels continued to be imported and were distributed widely. The six pieces of Roman glass vessels unearthed from the Eastern Jin Dynasty (317-420) tombs at Nanjing indicate that imported goods probably continued to arrive by sea to Southern Chinese ports. In northern China, however, trade must have relied on the overland route. The Western Jin Dynasty (265-316) poet Pan Ni, in an ode to a glass bowl, wrote about the difficult eastward journey of foreign glass, and he enumerated the stages of its dangerous passage through the scorching deserts and over the steep peaks of the Pamirs. This overland route, the Silk Road, could also be called the "Glass Road." Fragments of Roman glass found in many places in Xinjiang support this theory.

The glass vessels imported during the Sui and Tang Dynasties, excavated from Xian and Luoyang, show the two capitals' important position in trade between the East and the West. The fragments of Sasanian cut glass found in Xinjiang and the Sasanian glass bowl in the wall painting of Mogao Cave at Dunhuang are further evidence that Sasanian glass was imported via the Silk Road.

During the Song Dynasty (960-1279), Islamic glass was the main glass import. The discovery at the stupa No. 5 in Dingxian of eight vessels is an example of glass imports from the Caliphate. The records of the Song Dynasty frequently mention the importing of glass. In general, the places in China where foreign glass has been found indicate that the importation routes of glass vessels coincide with those of other goods.

Editor's Note: The author did not provide notes or references.

Figure 2. *Cup from an Eastern Jin tomb at Xiangshan, Nanjing, Jiangsu Province.*

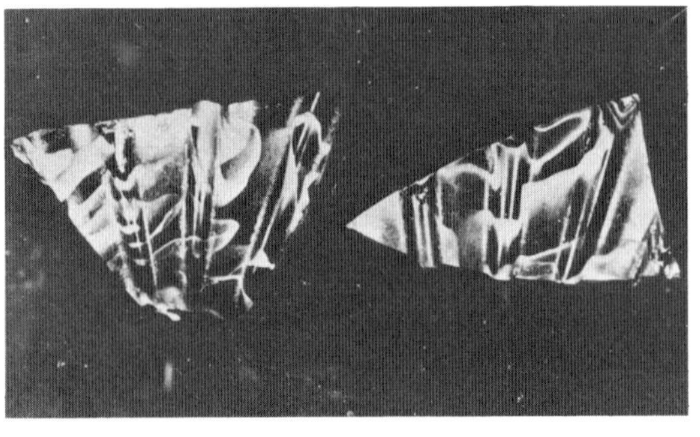

Figure 1. *Fragments of a ribbed bowl, from an Eastern Han tomb at Ganjiang, Jiangsu Province.*

Figure 3. *Cut bowl, from a Northern Zhou tomb at Guyuan, Ningxia.*

Figure 4. *Green bottle from a Tang tomb at Luoyang, Henan Province.*

Figure 6. *Engraved bottle from a Northern Song stupa at Wuwei, Anhui Province.*

Figure 5. *Engraved bottle from a Northern Song stupa at Dingxian, Hebei Province.*

Figure 7. *Engraved bottle from a Northern Song stupa at Ruian, Zhejiang Province.*

Figure 8. *Blue bottle from a Northern Song stupa at Dingxian, Hebei Province.*

Figure 9. *Green cup from the tomb of Geng Yanyi, Liao Dynasty, at Chaoyang, Liaoning Province.*

Figure 10. *Plate from the tomb of Liu Sheng, Western Han Dynasty, at Mancheng, Hebei Province.*

Figure 11. *Double-handled cup from the tomb of Liu Sheng.*

Figure 12. *Bowl and bottle from a Northern Wei stupa at Dingxian, Hebei Province.*

Figure 13. *Box and tube from the tomb of Li Jingxu Sui Dynasty, at Xian, Shaanxi Province.*

Figure 14. *Two egg-shaped vessels from the tomb of Li Jingxun.*

Figure 15. *Green bottle from the tomb of Li Jingxun.*

Figure 16. *Two small cups from the tomb of Li Jingxun.*

Figure 17. *Green bottle from the tomb of Li Tai, Tang Dynasty, at Yunxian, Hubei Province.*

Figure 18. *Small bottle from a Tang stupa at Ningan, Heilongjiang Province.*

Figure 19. *Gourd-shaped bottle from a Northern Song stupa at Lingtai, Gansu Province.*

Figure 20. *Green bowl from the Dingxian stupa No. 6, Northern Song Dynasty, Hebei Province.*

Table 1

Glass Vessels Excavated in China after 1949

No.	*Site*	*Date*	*Description*	*Source*
1	Tomb of Liu Shen at Mancheng, Hebei	Western Han 113 B.C.	Dish, 2 double-handled cups	Domestic
2	Tomb at Hengzhigang, Guangdong	Western Han	3 bowls	Roman ?
3	Tomb at Hepu, Guangxi	Western Han	Bowl	
4	Tomb No. 2 at Ganquan, Ganjiang, Jiangsu	Eastern Han 76 A.D.	Fragments of ribbed bowl	Roman
5	Tomb at station of Guixian, Guangxi	Eastern Han	Bowl	Roman ?
6	Tomb at Nandoucun, Guixian, Guangxi	Eastern Han	Cup with dish	Roman ?
7	Tomb at Guixian, Guangxi	Eastern Han	Bowl	
8	Tomb at Guixian, Guangxi	Eastern Han	Bowl	Roman
9	Tomb No. 121 at Wuli-dun, Echeng, Hubei	Eastern Han	Bowl	Sasanian
10	Tomb of Huafang at Beijing	Western Jin	Bowl	Sasanian
11	Tomb at Shimenkan, Nanjing	Southern Dynasties	Fragments	Roman
12	Tomb at Caidian, Hanyang, Hubei	Southern Dynasties	Bottle	
13	Tomb No. 7 at Xiang-shan, Nanjing	Eastern Jin 322 A.D.	2 cups	Roman
14	Tomb at Nanjing University	Eastern Jin	Cup	Roman
15	Tomb at north suburb of Nanjing	Eastern Jin	2 cups	Roman

16	Tomb of Feng Sufu at Beipiao, Liaoning	Northern Yan 415 A.D.	2 bowls, duck-shaped bottle	Roman
17	Stupa at Dingxian, Hebei	Northern Wei 481 A.D.	Bowl, 2 bottles 3 gourds	Domestic
18	Tomb of Feng Monu at Jingxian, Hebei	Northern Wei 521 A.D.	Bowl	Roman
19	Tomb of Zu Shi at Jingxian, Hebei	Northern Wei	Cup	Roman
20	Tomb at Jurong, Zhen-Jiang, Jiangsu	Southern Dynasties	Bowl	Sasanian
21	Tomb of Li Xian at Guyuan, Ninxia	Northern Zhou	Bowl	Sasanian
22	Tomb No. 586 at suburb of Xian, Shaanxi	Sui 592 A.D.	Stemmed cup	
23	Stupa at Yaoxian, Shaanxi	Sui 604 A.D.	Bottle	Domestic
24	Tomb of Li Jingxin at Xian, Shaanxi	Sui 608 A.D.	2 bottles, 2 cups 2 eggs, tube	Domestic
25	Tomb of Ji Wei at Xian, Shaanxi	Sui 610 A.D.	Small jar, 2 cups	Domestic
26	Tomb No. 1 at Quinzhou, Guangxi	Sui	Cup	
27	Tomb of Li Shou at Sanyuan, Shaanxi	Tang 631 A.D.	3 bottles	Domestic
28	Tomb of Li Tai at Yunxian, Hubei	Tang 652 A.D.	Cup, 3 bottles	Domestic
29	Tomb of Li Shuang at Yangtouzhen, Shaanxi	Tang 668 A.D.	Cup	
30	Stupa at Jinchuan, Gansu	Tang 694 A.D.	Bottle	Domestic
31	Stupa at eastern sub-urb of Xian, Shaanxi	Tang 719 A.D.	Bottle	Domestic
32	Stupa at Ningan, Heilongjiang	Tang	Bottle	Domestic

33	Tomb No. 118 at Guanlin, Luoyang, Henan	Tang	Bottle	Sasanian
34	Hoard at Hejiacun, Xian, Shaanxi	Tang	Cup	Sasanian
35	Site of west market, Xian, Shaanxi	Tang	Fragments	
36	Tomb No. 504 in suburb, Xian, Shaanxi	Tang	Fragments	
37	Stupa No. 5 at Dingxian, Hebei	Northern Song, 976 A.D.	7 bottles, 3 bowls 2 cups, 8 gourds	Islamic
38	Stupa No. 6 at Dingxian, Hebei	Northern Song, 995 A.D.	33 gourds, bowl	Domestic
39	Stupa at Mixian, Henan	Northern Song, 999 A.D.	About 50 pieces of vessels	Domestic
40	Stupa at Xian, Shaanxi	Northern Song	Bowl	
41	Stupa at Lingtai, Gansu	Northern Song	3 gourds	Domestic
42	Stupa at Lianyuangang, Jiangsu	Northern Song 1027 A.D.	3 gourds	Domestic
43	Stupa at Ruian, Zhejiang	Northern Song, 1034 A.D.	Bottle Small bottle	Islamic Domestic
44	Stupa at Wuwei, Anhui	Northern Song 1036 A.D.	Bottle	Islamic
45	Stupa at Zhenjiang, Jiangsu	Northern Song 1078 A.D.	Bottle	Domestic
46	Stupa at Shouxian, Anhui	Song	2 small bottles	Domestic
47	Tomb at Dayingzi, Chaoyang, Liaoning	Liao	Bottle	
48	Tomb at Yemotai, Faku, Liaoning	Liao	Square plate	
49	Tomb of Gengyanyi at Chaoyang, Liaoning	Liao 1020 A.D.	Handled cup, dish	Islamic
50	Tomb at Yixian, Liaoning	Liao	Bowl	

Table 2

Quantitative Chemical Analyses

No.	Description	Date	SiO_2	Al_2O_3	CaO	Na_2O	K_2O	MgO	Fe_2O_3	MnO_2	CuO	PbO	Other
4	Ribbed bowl	Han	64.79	3.44	7.66	18.18	0.88	0.61	1.30	2.45	0.03		
	Ear ornament	Han	49.33	2.10	3.16	9.30	0.51	1.40	0.48	0.33	0.09	21.62	BaO: 8:83 CoO: 0.04
9	Cut bowl	Jin	64.22	1.64	9.19	17.51	3.59	3.21	0.57	0.04	0.02		
14	Cut cup	Jin	69.39	1.89	6.81	19.60	0.49	0.27					Sb_2O_3 0.29
15	Cut cup	Jin	67.70	3.43	6.05	19.23	0.45	0.94	0.58	1.63	0.02		
15	Vessel	Jin	69.15	2.09	5.79	15.84	0.33	0.55	1.22	0.03	0.27		CoO
16	Bowl	Yan	64.82	2.71	6.14	16.02	4.43	2.35	0.82	0.08	0.02		
27	Bottle	Tang	36.16	2.42	1.09	10.01	0.95	2.84				46.65	
28	Bottle	Tang	61.58	1.66	6.27	17.86	3.53	6.43	0.69				
28	Bottle	Tang	30.49	1.61	0.20	0.31	0.27	0.30	0.33			64.23	
33	Bottle	Tang	62.70	2.75	6.57	18.11	2.60	4.84	1.05	0.04			TiO_2, 0.15
	Bead	Tang	26.08		0.18	0.29	0.06	0.09	0.26	0.02	0.41	68.51	
37	Gourd	Song	26.85		0.35	0.18	0.34	0.10	0.19	0.02	0.41	70.04	
41	Gourd	Song	36.32		0.13	0.29	10.09	0.10	0.16	0.02	0.13	50.31	
41	Gourd	Song			0.18	0.11	11.94	0.04	0.29	0.01		53.40	
39	Egg	Song	31.66	2.22	3.35	0.11	13.75	0.30	4.39			41.57	
	Xinjiang	?	67.24	2.58	9.18	18.31	0.85	0.54	0.30	0.12	0.02		
	Xinjiang	?	69.89	1.98	5.77	19.18	0.48	0.52	0.31	0.03	0.02		
	Xinjiang	?	54.14	7.78	9.11	15.66	5.66	5.50	1.06				

Xinjiang	Yuan	60.43	4.09	8.16	15.87	5.27	4.71	0.35	0.04	0.02
Xinjiang	Yuan	58.27	5.38	11.88	12.87	5.51	4.04	0.97		
Xinjiang	Yuan	61.03		5.68	15.18	6.06	3.81	1.01	0.04	

Table 3

Qualitative XRF Analyses

No.	Description	Date	Si	Cl	K	Ca	Mn	Fe	Ni	Cu	Sr	Zr	Pb	Ag	In	Sn	Ba	Sb
2	Blue bowl	Han	+++		++	++		+		+	+		+				+	
8	Green bowl	Han	+++		++	+++		+									+	
18	Green bowl	Wei	+++	++	++	+++		+		+	+	+	+	+	+	+		
19	Greenish cup	Wei	+++	++	++	+++	+	+		+	+			+	+			
24	Green bottle	Sui	+++		++	+++		+	+								+	
24	Green box	Sui	+++		++	++							+++					
24	Green box (lid)	Sui	+++		++	++								+++				
24	Light blue cup	Sui	+++		++	+++		+					+			+		
24	Light blue cup	Sui	+++		++	+++		+								+	+	
24	Greenish cup	Sui	+++	++	++	+++		+									+	
35	Fragment	Tang	+++	++	++	+++	+	+										
	Yellow figure	Tang				++				+			+++					
37	Sky-blue bottle	Song	+++	+	++	++		++		++			++					

CHAPTER 2

A "GLASS GARMENT" FROM A WESTERN HAN TOMB IN JIANGSU PROVINCE

CHENG ZHUHAI AND ZHOU CHANGYUAN

Research Institute of Building Materials, Beijing;
Yangzhou Museum, Yangzhou, Jiangsu

IN 1977, archaeologists of the Yangzhou Museum unearthed a tomb at Ganquanshan in Yangzhou in which was a lady of noble rank named Mo Shu. The excavation and related finds revealed that this was one of the Liu family tombs dating to the late Western Han Dynasty (206 B.C - 24 A.D.)[1]

A "glass garment," one of many types of funeral objects and adornments discovered in the tomb, had never before been found in China. According to these studies, the pieces from the Han tomb in Yangzhou are clearly made of glass and we have named it *glass fabric*.[2] This adds to historical records pertaining to the special funeral garments used in the Han Dynasty; only jade and stone garments are mentioned in the documents of that period.

We believe that all the glass garment pieces were mold-pressed. The pieces of each kind are of identical size, especially the decorated pieces. This fabric appears to have been a sort of adornment for the head or body, and it consists of about 600 pieces of glass of different sizes and shapes. There are eleven shapes in all (see Fig. 1 and Table 1), including rectangles, ladders [or trapezoids], and circles. The rectangular pieces (6.2 x 4 cm) are predominant; the smallest piece is narrow and ladder-shaped (6.2 x 1 x 2 cm). All of the glass pieces are $\sim$ 0.4 cm thick.

The round pieces and a few of the rectangular ones have surface decorations of hydras [octopi?], flowers, and plants while other pieces have undecorated surfaces (Figs. 2, 3). In the middle of the decoration on some rectangular pieces, there is a four-petaled flower, and some of these flowers have the remains of gold foil on the pistils. All of the round, multi-angular, and narrow ladder-shaped glass pieces are pierced with three holes, while the other pieces are pierced with four.

Because there is a gray surface on the pieces of glass, the glassy layer (Fig. 4) can only be seen when viewed in cross-section, and for this reason the glass has often been mistaken for stone or ceramic. With x-ray studies, the diffraction patterns show glassy characteristics for the inner layer but indicate that crystalline phases are present in the outer, gray surface layer.

The results of our chemical analyses are reported in Table 2. It can be seen that the glass composition is in the PbO-BaO-SiO_2 system. In accordance with the x-ray analyses, the pieces also show glass properties by petrographic examination (Fig. 5). The refractive index of the glass is 1.612 $\pm$ 0.0030. Some crystal inclusions show indices 0.001-0.005 lower than the glass. A cross-section of a piece was sampled for scanning electron microscopic examination, and the crystal inclusions were analyzed by EDAX (Figs. 6 and 7). The results show that these crystals are rich in SiO_2, PbO, and BaO. Perhaps they are crystals of lead-barium silicate or barium silicate.

We also observed the behavior of the pieces

under heat treatment. For the inner layer, the hemisphere point is 870°C, and the fluid point is 933°C. The hemisphere point of the surface layer is 1156°C. No sphere points were observed for either case.

It is likely that before the mold pressing, the glass batch powder was spread on the bottom of the ceramic mold uniformly. After the casting of glass melts, the glass batch powder was again spread on the surfaces of still-softened glass before it was mold-pressed. Thus the two layers joined tightly. Furthermore, the surface layers are of different thicknesses, ranging from 0.1 to 0.4 mm and the gold foil still remained in the pistil decoration on the surface. All these observations would be impossible if the surfaces had been heavily weathered. It is likely that the craftsmen spread glass batch powder on the surfaces of the ceramic molds to facilitate the drawing of patterns. {Editor's Note: An alternative view is that the gray surface layer consists of weathering products and/or contains surface-nucleated devitrification phases.}

For several reasons, we believe that the glass making up the garment was independently made in China. China had developed glassmaking at least as early as the Warring States Period, according to excavation findings and historical documents from the Han Dynasty. The hydra, flower, and plant decorations were all very popular in China at that time. Compositionally, the glass is of the $PbO\text{-}BaO\text{-}SiO_2$ system, which is obviously different from that of the Western World.[3,4]

The discovery of this glass garment in Yangzhou suggests that glassmaking, especially mold pressing, had developed to a high level at the time of the Han Dynasty. The decorations on the glasses, and the addition of gold foil, indicate the outstanding glassmaking skills possessed by ancient Chinese craftsmen.

Notes

1. Yangzhou (or Nanjing?) Museum, "A Western Han Tomb: The Concubine of Mo Su," *Wenwu*, 1980, no. 2, p. 1. (In Chinese.)
2. The Chinese term does not seem to translate smoothly into English but we will use throughout, and interchangeably, the terms *glass fabric* or *glass garment* according to which reads better in context.
3. Cheng Zhuhai, "Preliminary Research on the Development of Ancient Chinese Glasses," *Journal of the Chinese Silicate Society*, v. 9, no. 1, 1981, pp. 96-104.
4. Zhang Fukang, Cheng Zhuhai, and Zhang Zhigang, "An Investigation of Ancient Chinese *Liuli*," *Journal of the Silicate Ceramic Society*, v. 11, no. 1, 1983, pp. 67-76.

Table 1

Classification of "Glass Cloth" Pieces According to Shapes

	Shape	*Length x width or length x upper x lower width*	*Thickness*	*Number of holes*	*Decoration*
1	rectangular	6.2 x 4 cm	0.4 cm	4	without
2	rectangular	6.2 x 3.8	0.4	4	without
3	narrow ladder-shaped	6.4 x 1.7 x 1.4	0.4	4	without
4	ladder-shaped	6.2 x 2.5 x 2.1	0.4	4	without
5	narrow ladder-shaped	6.5 x 1.2 x 1	0.4	3	without
6	multi-angular	6.2 x 0.8 x 4	0.4	3	without
7	narrow ladder-shaped	6.4 x 2 x 1.8	0.4	4	without
8	ladder-shaped	6.2 x 3.5 x 3	0.4	4	without
9	rectangular	6.2 x 3.6	0.4	4	without
10	rectangular	6.2 x 4	0.4	4	with
11	round	diameter 6.2	0.4	3	with

Table 2

Chemical Analyses of the "Glass Fabric"

	Interior glass	*Surface layer*
SiO_2	36.03	30.27
PbO	40.37	50.39
BaO	21.49	3.60
Al_2O_3	0.02	0.12
K_2O	0.07	0.04
Na_2O	2.27	2.28
CaO	0.22	0.70
MgO	0.08	0.03
Fe_2O_3	0.07	0.32
Total	100.62	87.75

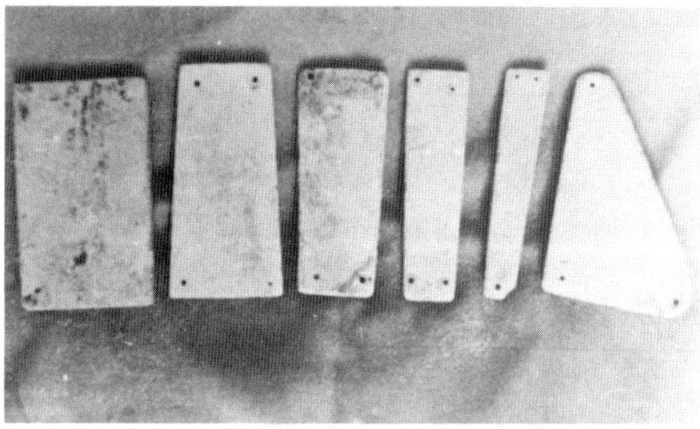

Figure 1. *Trapezoidal pieces from glass garment.*

Figure 3. *Rectangular pieces with hydra and floral decorations.*

Figure 2. *Round pieces with decoration.*

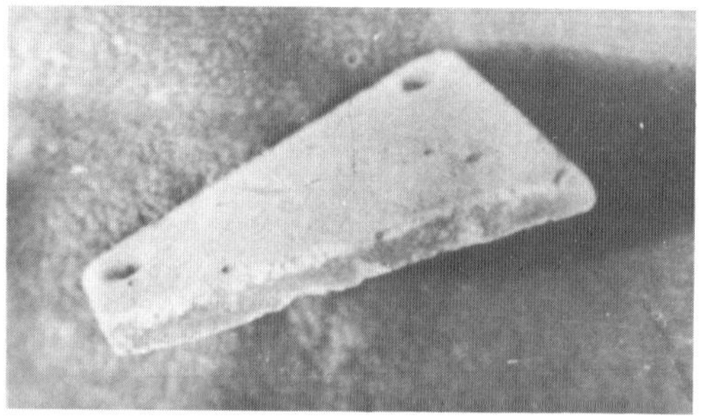

Figure 4. *Fractured section of rectangular piece.*

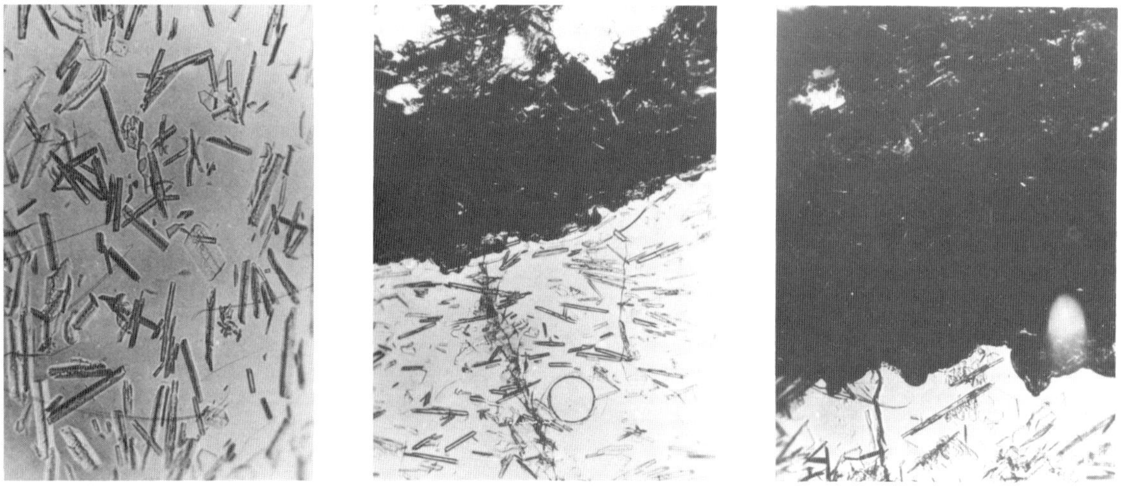

Figure 5. *Three petrographic sections.*

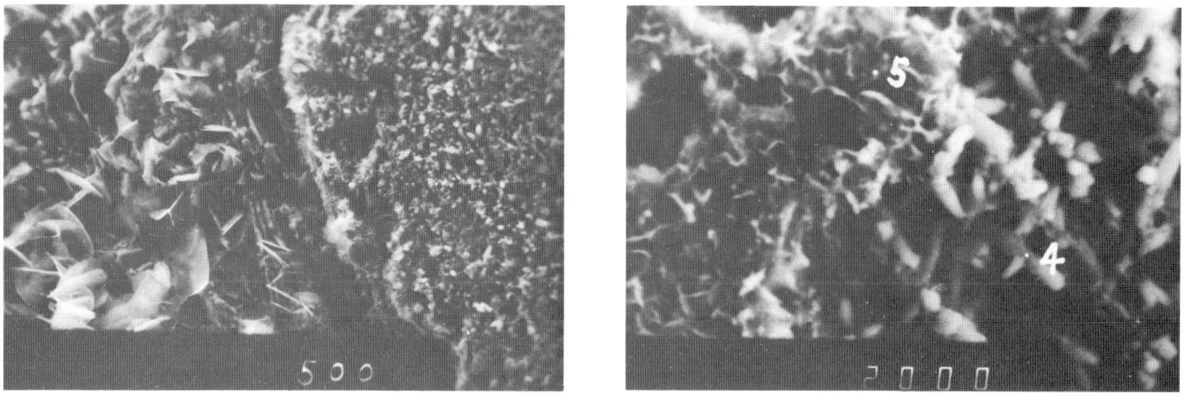

Figure 6. *SEM photographs.*

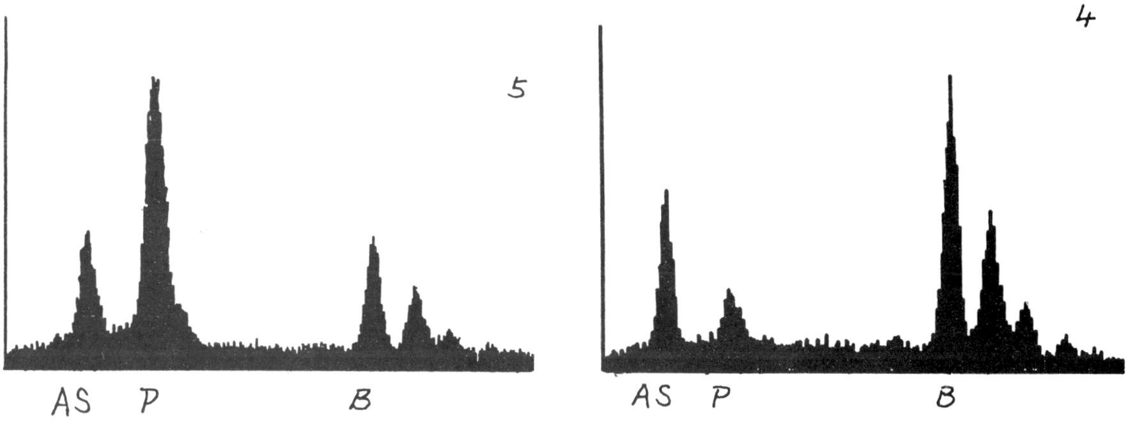

Figure 7. *EDAX results of glass garment analysis.*

CHAPTER 3

INVESTIGATIONS OF SOME ANCIENT CHINESE LEAD GLASSES

SHI MEIGUANG, HE OULI, WU ZONGDAO, AND ZHOU FUZHENG

China Building Materials Academy,
Beijing

EIGHTEEN examples of ancient Chinese lead glass have been examined and analyzed. The glasses were excavated in ten places located in six provinces of China and date from the Warring States Period (475-221 B.C.) to 999 A.D. in the Song Dynasty; the dates and localities of excavation were provided by archaeologists. The earliest lead glass sample belongs to the Warring States Period and the latest to the Song Dynasty.

The excavation sites were in Hunan, Henan, Hebei, Gansu, Shaanxi, and Liaoning. The eighteen pieces were chosen as being representative of lead glasses excavated in China during the past three decades, and the samples are described in the Catalogue of Samples. The chemical compositions fall into three groups consisting of five systems. From the findings of these investigations, the authors believe that these glasses were made in China.

Most lead glass objects of the Warring States Period and Han Dynasty, of which there are only few examples, are ornamental. These early objects include ritual disks, beads, and earrings, the ritual disks being 80-140 mm in diameter and 3-5 mm thick with exquisite decorative patterns on the surfaces. However, the majority of lead glass objects of the later Tang and Song Dynasties are bottles and calabashes in addition to some beads, eggs, and a goose. The heights and diameters of these Tang and Song objects are generally no more than 100 mm.

As to color, many of the ancient Chinese lead glasses are green owing to the presence of copper and iron, but there are also some white opaque, amber, blue-violet, red, and dark brown glasses, indicating that during this historical period Chinese craftsmen had mastered the art of glass coloring.

Lead glasses of the Warring States Period and Han Dynasty were made by mold pressing and casting; there was no glassblowing. During the Tang Dynasty, lead glass was blown and cast into figurines. From the many blown glass objects of the Song Dynasty, such as the glass goose excavated in Mixian (GS-20, Fig. 1), it can be seen that by the 10th century Chinese craftsmen had learned to blow glass.

The samples were examined by means of a polarizing microscope and X-ray diffraction. All proved to be made of glass. Using densitometry, refractometry, and spectrometry, the samples in Table 1 were confirmed to be lead glasses. These samples were examined after removing the corrosion products, and they were found to be optically isotropic and contained some seed (minute bubbles), confirming that they are glasses. A great number of minute seed 3-10 microns in diameter were observed in the milky-white ritual disks GW-1 and GW-2 (sample thickness 0.1 mm). Although many lead-barium opal glasses of the Warring States Period have been excavated, the source of the opacity has not been studied before. We believe it results from the dispersion of light caused by the seed. Sample GW-1 was also examined by X-ray diffraction. No characteristic lines of crystal phases were observed. This result is consistent with the microscopic observation and with our conclusion that the opacity is due solely to seed.

The corrosion products of the green glass beads GT-9 (Fig. 2) were also studied by X-ray diffraction. The results show that the outer layers of the PbO-SiO_2 glass with 68.51% PbO had become crystalline lead carbonate, due to the weathering process.

Thirteen pieces of lead glass were analyzed chemically, and the results are reported in Table 1. One of the samples, GT-8, was analyzed quantitatively by EDAX, the others by atomic absorption and wet chemical methods. The corrosion products were removed before analysis. The compositions of ancient Chinese lead glasses can be classified into three groups consisting of five systems and these are:

1. *Lead-Barium-Silica Glasses*
   Samples GW-1, GW-2 and GW-3 can be assigned to the PbO-BaO-SiO_2 system. GH-6, with 9.3% Na_2O, belongs to the Na_2O-PbO-BaO-SiO_2 system. The lead-barium glasses analyzed here contain 21-45% PbO and 10-13.4% BaO.
2. *Lead-Silica Glasses*
   Samples GT-9 and GS-14, with high PbO contents, can be assigned to this system. GS-14 contains 70% PbO.
3. *Alkali-Lead-Silica Glasses*
   Samples GS-11, GS-15, GS-18, GS-19, GS-20, and GS-24 can be assigned to the K_2O-PbO-SiO_2 system, with 6-15% K_2O. GT-8 belongs to the Na_2O-PbO-SiO_2 system. In the tenth century, glasses of the K_2O-PbO-SiO_2 system seemed to be common in China. We have found them not only in Mixian, Dingxian, and Lingtai, but also in a small bottle of this system which was transported from China to Japan in the 10th century.[1]

The Chinese lead-barium composition was unique among the glasses of the ancient world. In addition, the ancient Chinese lead glass is not only abundant and widely dispersed, but also varied somewhat within its compositional system. Ancient lead glass was much less common in the West.[2] {Editor's Note: "Lead glasses" as such really did not exist at all in the ancient West, in the sense that lead oxide was never a major ingredient of a base glass composition. Instead, it found use only — although extensive use — in glass colorants, namely, in yellow opaques as Pb_2SbO_7 and $PbSnO_3$, and in red opaques. In the West, the earliest presently-known uses of lead as a major ingredient in a base composition was in emerald green Islamic cameo glasses of the 10-11th centuries and in certain Eastern European glasses, most often in the form of beads.}

As early as the Western Zhou Dynasty (1066-770 B.C.), the ancient Chinese could make faience beads by sintering quartz powder, and they had acquired knowledge of the usefulness of lead compounds for glassmaking during the Warring States Period. During the Western Zhou Dynasty, the firing temperature of ceramics had reached 1200°C, high enough for glass melting, which made it quite possible to create lead glasses in China.

Acknowledgments. The authors are grateful to the Research Institute of Archaeology, the Cultural House of Mixian, the Museum of Hunan Province, and the Museum of Dingxian for providing the glass samples.

Notes

1. K. Yamasaki, "*Chemical Compositions of Ancient Glasses found in Japan*," Tenth International Congress on Glass, No. 9, The Ceramic Society of Japan, 1974, pp. 15-20.
2. Earle R. Caley, *Analyses of Ancient Glass, 1790-1957*, The Corning Museum of Glass, 1962.
3. Museum of Shaanxi Province, *Wenwu*, 9, 1974, p. 77. (In Chinese.)
4. Qin Ming-zhi, *et al.*, *Wenwu*, 2, 1983, p. 39. (In Chinese.)
5. Museum of Dingxian, *Wenwu*, 8, 1972, p. 39. (In Chinese.)
6. Jin Ge, *Wenwu*, 10, 1972, p. 63. (In Chinese.)

Catalogue of Samples

GW-1 Ritual disk, Warring States; Changsha, Hunan. Mold pressed, carved. Milky white; 3.89 g/cc.

GW-2 Ritual disk, Warring States; Hengyang, Hunan. Mold pressed, carved, patterned. White opaque; 3.95 g/cc.

GW-3 Ritual disk, Warring States; Changsha, Hunan. Mold pressed, carved. Green.

GH-5 Earring, Western Han; Yuxian, Henan. Dark green, heavily weathered. H. 12 mm.

GH-6 Earring, Han; Jiauquan, Gansu. Blue-violet; 3.25 g/cc. Trumpet-shaped terminal.

GH-7 Beads, Eastern Han; Changde, Hunan. Cast. Bright green, heavily-weathered. $n > 1.698$. D. 10 mm.

GT-8 Bottle, 631 A.D.; Sanyuang, Shaanxi. Blown. Green, heavily-weathered. Th. 0.9 mm. Ref. 3.

GT-9 Bead, 691 A.D.; Chaoyang, Liaoning. Cast. Bright green, heavily-weathered; 5.31 g/cc. D. 9 mm.

GS-10 Figurine, 691 A.D.; Chaoyang, Liaoning. Mold cast. Green, heavily-weathered; 5.09 g/cc.

GS-11 Bottle, 10th century; Lingtai, Gansu. Blown. Greenish. H. 7.8 cm. Ref. 3.

GS-13 Bottle, 10th century; Lingtai, Gansu. Blown. Green. $n = 1.655$. H. 98 mm. Ref. 4.

GS-14 Glass calabash, 977 A.D.; Dingxian, Hebei. Blown. Green; 5.34 g/cc. H. 50 mm. Ref. 5.

GS-15 Glass grapes, 977 A.D.; Dingxian, Hebei. Dark brown, hollow in center; 3.75 g/cc. Ref. 5.

GS-17 Bowl, 995 A.D.; Dingxian, Hebei. Green, lobed rim; 3.90 g/cc. D. 15.8 cm, H. 9.0 cm. Ref. 5.

GS-18 Glass egg, 999 A.D.; Mixian, Henan. Blown. Dark red, heavily-weathered. L. 5.5 cm. Ref. 6.

GS-19 Glass egg. Same as GS-18, but dark yellow.

GS-20 Glass goose, 999 A.D.; Mixian, Henan. Blown. Green. H. 5.5 cm.

GS-24 Glass calabash, 999 A.D.; Mixian, Henan. Blown. Green; 4.26 g/cc.

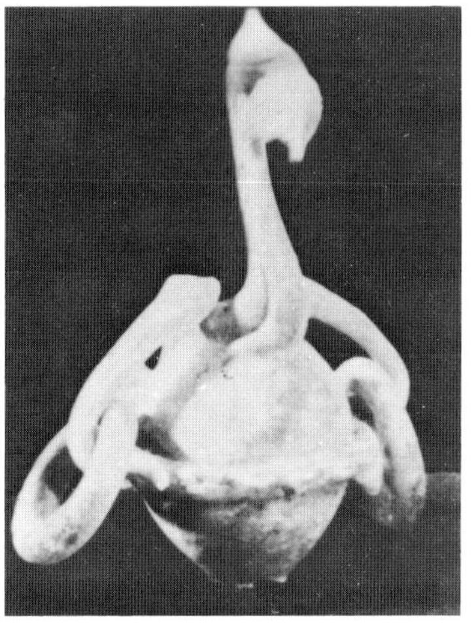

Figure 1. *Glass goose (GS-20, 999 A.D.).*

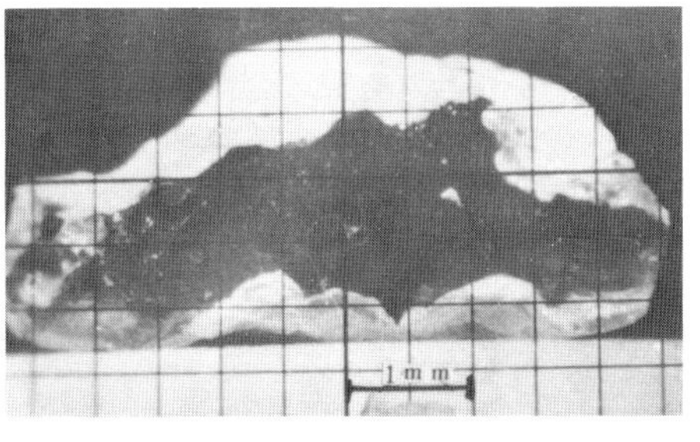

Figure 2. *Cross-section of lead glass bead (GT-9, 691 A.D.). Shows heavily-weathered surfaces with unweathered interior.*

Table 1

Chemical Analyses of Some Ancient Chinese Lead Glasses

	GW-1	*GW-2*	*GW-3*	*GW-6*	*GT-8***	*GT-9*	*GS-11*	*GS-14*	*GS-15*	*GS-18*	*GS-19*	*GS-20*	*GS-24*
SiO_2	37.16	36.57	38.30	49.33	36.16	26.08	36.32	26.85	36.93	33.78	31.66		30.02
Al_2O_3	0.62	0.46	1.67	1.42	2.42				1.11	2.62	2.22		0.62
Fe_2O_3	0.16	0.15	0.22	0.48		0.26	0.16	0.19	4.13	3.15	4.39	0.15	0.28
PbO	39.80	44.71	41.53	21.62	46.65	68.51	50.31	70.04	45.93	40.15	41.57	47.34	57.25
BaO	13.4	10.1	10.4	10.5									
CaO	1.95	2.1	2.75	3.16	1.09	0.18	0.13	0.35	0.36	3.52	3.35	0.17	2.32
MgO	0.40	0.21	0.41	1.40	2.84	0.09	0.1	0.1	0.08	0.31	0.3	0.04	2.26
K_2O	0.27	0.10	0.34	0.51	0.95	0.06	10.09	0.34	8.45	14.78	13.75	11.45	6.08
Na_2	3.32	3.72	3.07	9.30	10.01	0.29	0.29	0.18	0.08	0.13	0.11	0.08	0.16
CuO	0.03	0.02	0.73	0.09		0.41	0.13		1.44	1.32	0.4	0.18	1.0

CHAPTER 4

CHEMICAL ANALYSES OF SOME EARLY CHINESE GLASSES

ROBERT H. BRILL, STEPHEN S. C. TONG, AND DORIS DOHRENWEND

The Corning Museum of Glass and Corning Glass Works,
Corning, N.Y.; and The Royal Ontario Museum, Toronto

COMPARED to what is known about the history of glass made in "the Western World,"[1] our knowledge of early Chinese glass is really quite sparse.[2] The Introduction to this book lists more than a dozen truly fundamental questions about the history of glass and glassmaking in China which still remain either unanswered or only partially answered.[3] In this paper, and two companion papers, scientific evidence is presented which we hope will be useful in answering some of these questions. The present paper is concerned primarily with chemical analyses. In all, seventy-one samples of Chinese glasses and related materials have been studied.

The analytical results are most relevant to questions such as these:

1. How can one distinguish between glass made in China and glass imported from elsewhere?
2. Was glassmaking discovered in China independently from its discovery in the Near East? (Glass first appeared in Egypt and Mesopotamia about 1500 B.C.)
3. What materials were used for making glass in early times in China, and where did they come from?
4. What chemical formulations were used in different parts of China and during different historical periods, and what can be inferred from any differences in those formulations?
5. What accounts for the presence of barium in so many early Chinese glasses?
6. How was glass technology in China related to early metallurgy and to the manufacture of faience, glassy faience, "Chinese blue," glazed ceramics, etc.?

The main obstacle to analyzing Chinese glasses has been the limited availability of samples suitable for analysis. Most of the collections of Chinese glass in Western museums contain only intact objects which would be disfigured by the removal of samples large enough for quantitative analysis. Moreover, most of the Chinese glasses available for analysis in the West are without provenance; even after analyses are completed one cannot be certain of exactly what periods and places of manufacture they represent.

Analytical Results

Quantitative chemical analyses have been carried out on forty-nine specimens of early Chinese glasses using a combination of atomic absorption, emission spectrography, and the electron microprobe. Twenty-six of the samples came from fragments and objects in the Royal Ontario Museum, eighteen from objects in The Corning Museum of Glass, two from the Beck Collection, one from the Museum für Kunst und Gewerbe, and three from private collections. These objects all entered their respective collections many years ago, and the majority are of unknown provenance. The dates and descriptions in the Catalogue of Samples at the end of this article were provided by one of the authors (D.D.), and by the late Dorothy Blair who catalogued the East Asian glass in The Corning Museum of Glass. For simplification, the samples have been separated arbitrarily into three broad chronological groups, although it is recognized that wide variations in composition and technology may exist within each. The groups are: Early Periods (ca. Warring States through Sui; ca. 5th century B.C. - 7th century A.D.), Middle Periods (ca. Tang through Yuan; ca. 7th-14th centuries), and Late Periods (ca. Ming and following). Most of the objects studied are illustrated in Figures 1-18.

The analyses were carried out on samples weighing less than 50 mg, and the remainders of the samples were used for the lead isotope determinations described in a companion paper. It was agreed by all concerned that the sacrifice of the minute amounts of material required was justified by the information to be gained from the analyses. In a few cases the samples were too small for quantitative analyses and only semi-quantitative spectrographic analyses were run.

The analyses revealed five basic types of compositions which (with some variants) separate the objects roughly according to their periods of manufacture. Similar compositions for Chinese glasses, from about the same time periods, have been reported or alluded to previously.[4] The actual analyses are presented in Tables 1-3, and summarized in Table 4, where the various compositional types are shown along with the dates of the glasses analyzed. Future analyses can be expected to expand and refine this picture considerably.

Interpretation of Results

A proper interpretation of this kind of analytical data can only be made in the light of an historical and cultural understanding of the contexts in which the glass was made, traded, and used. Therefore, we regard the analyses reported here as a starting point — a contribution towards a fuller understanding of early Chinese glass which all of us at this symposium look forward to in the future.

For the present, the glasses have simply been classified on the basis of compositional similarities and differences, as shown in Table 4. Glasses which are very similar in this table almost certainly bear some relationship in time and/or place of manufacture, or, in raw materials, formulations, and glassmaking practices. The classifications were drawn from graphical comparisons between single oxide concentrations, pairs of oxides, and various mathematical functions of groups of oxides. The graphs were examined for clusters of similar samples and for trends which indicate either positive correlations or substitutional effects. In the case of the high-lead glasses of the Early periods, various oxides or groups of oxides were subtracted out and the "balances" of the compositions normalized to 100%, after which the balances were treated as above. Several triangular diagrams were also plotted.

Even after much thought and numerous calculations, no simple and straightforward picture emerged as to the basic formulation of most of the glasses; nor was it possible to identify with confidence any of the primary batch materials. This could be because the glasses were made at different times, because the glasses were made from materials which were not uniform in composition, because some of the glasses might have been remelted and reworked, or because the glassmaking practices were not standardized.

It should be possible to clarify the picture eventually as additional analyses are completed and further methods of treating the data are developed. It should also be helpful to evaluate the chemical analytical data in conjunction with lead isotope data.

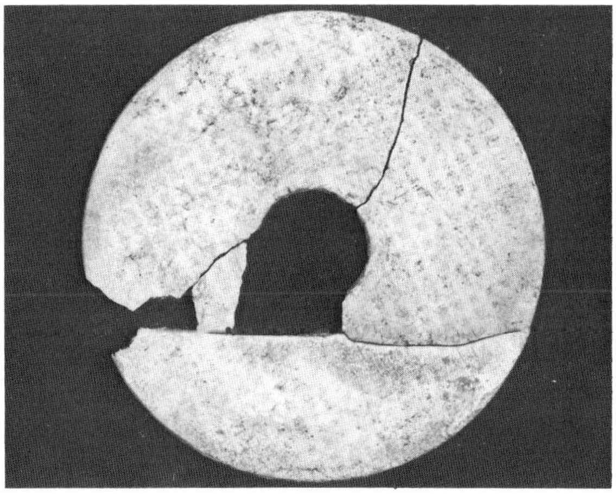

Figure 1. *Samples 05, 06*

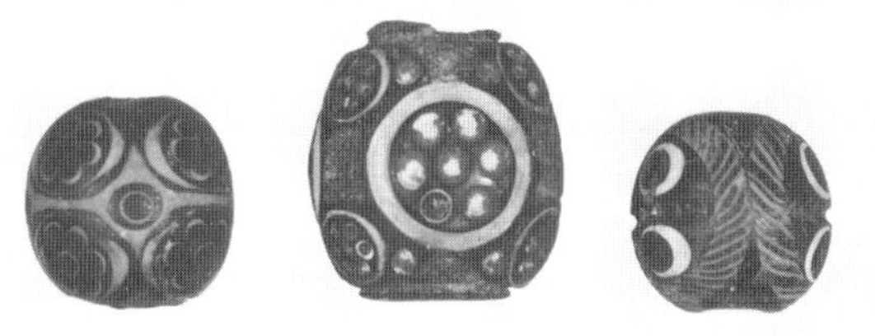

Figure 2. *Left to right, samples 52, 54, 50*

Figure 3. *Sample 09*

Figure 4. *Left, sample 80; upper right 83; lower right 84*

Figure 5. *Top, left to right, samples 73, 89, 92, 95; bottom 86, 91, 93, 94*

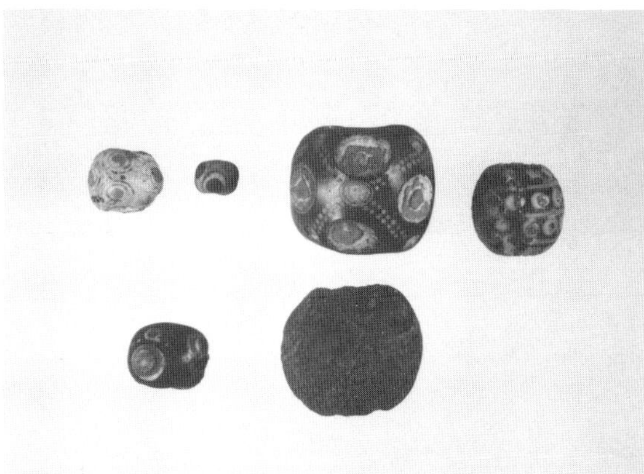

Figure 6. *Top, left to right, samples 76, 5817, 78, 79, & 5821; bottom 5823, 77*

Early periods, 5th cent. B.C.-7th cent. A.D.: High-lead glasses with barium ($PbO:BaO:SiO_2$)

The most interesting of all early Chinese glasses, chemically, are those having high PbO contents accompanied by high BaO. Such glasses have been noted by several authors previously,[5] and eighteen more examples are reported here. (See Table 4.) These glasses were unquestionably made in China because compositions of this sort are completely unknown elsewhere until the 19th-century. While certain ancient glasses from the West contain high concentrations of PbO, they never have such high BaO contents.

The barium could have been introduced intentionally as a separate ingredient, intended to perform either the fluxing function of alkali or the stabilizing function of lime. On the other hand, it might have been introduced because it was associated with the lead-containing ingredient. The barium also produced a certain turbidity in the glasses which might have been regarded as desirable by early Chinese glassmakers since it gave the glasses a jade-like appearance. Our electron microprobe analyses showed that the turbidity is due to the presence of barium disilicate devitrification crystals (Fig. 19). This was confirmed by x-ray diffraction.[6] The microprobe analyses showed that no. 80 contains 55.9% BaO and 44.3% SiO_2 and that no. 83 contains 56.2% BaO and 43.8% SiO_2. These values are in excellent agreement with the theoretical composition of barium disilicate: 56.1% BaO and 43.9% SiO_2.

Previously, one of the authors (R.H.B.)[7] had observed that the BaO content was directly proportional to the PbO content in the first few samples analyzed in Corning. Not only were they directly proportional, but also the percentages corresponded very closely to a 1:1 mole ratio of the two oxides, suggesting that the lead and barium were introduced simultaneously as a chemical compound, that is, some mineral containing equal molar proportions of the two elements. However, now that we have analyzed a greater number of glasses, that simple relationship does not seem to hold up as a generalization, the lead often being in excess of the 1:1 ratio.

The combined percentages of PbO and BaO among seventeen of these glasses ranges from 24 to 60% of the total glass composition (No. 54 is unusual in that it contains only 4.11%.). With only one exception (no. 86) there is more PbO than BaO on a weight percentage basis. Leaving out one extremely high value (no. 69), the PbO:BaO weight ratio ranges from 0.68 to 4.87. In trying to back-calculate the batch ingredients and "recipes" then, one should look for a mixture of materials which would yield an oxide ratio of approximately these values. On a molar basis, the PbO:BaO ratio varies from 0.71 to 3.35, but there are nine glasses having ratios between 0.859 and 1.27. These samples (nos. 05, 50, 52, 73, 79, 83, 84, 89, and 91) are close enough to a 1:1 ratio to keep open the possibility that these two elements might have been introduced as a compound.

There is another point to remember in connection with the barium. The most familiar naturally-occurring form of barium is the mineral barite, which is barium sulfate. But barium sulfate is a poor batch material for making glass. In 1977, with the help of the late Dr. August A. Erickson, one of the authors (R.H.B.) attempted to melt a glass duplicating an ancient Chinese composition: 45% PbO:14% BaO:38% SiO_2:3% Na_2O. Using barium sulfate as an ingredient, a poorly melted glass filled with much unreacted barium sulfate resulted, even after being soaked at 1200°C for 6 hours. On the other hand, using barium carbonate produced a well melted, transparent, homogeneous glass. Therefore, we believe that early Chinese glassmakers, instead of using barite as a raw material, are much more likely to have used witherite, the mineral form of barium carbonate. (Alternatively, they could

Figure 7. *Left to right, samples 82, 90, 1587, 1588*

Figure 8. *Left, sample 74, right 96*

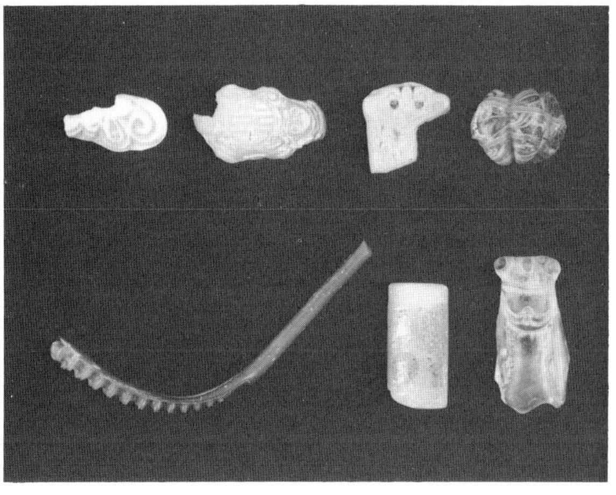

Figure 9. *Top, left to right, samples 5839, 1585, 5841, 5844; bottom, 1581, 75, 5805*

Figure 10. *Sample 3343*

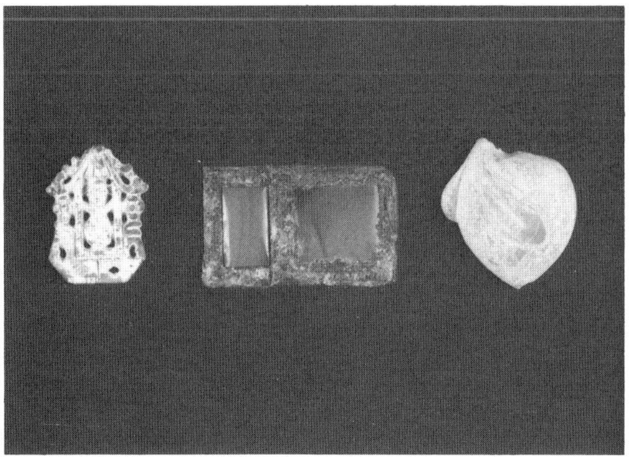

Figure 11. *Left to right, samples 5803, 72, 5811*

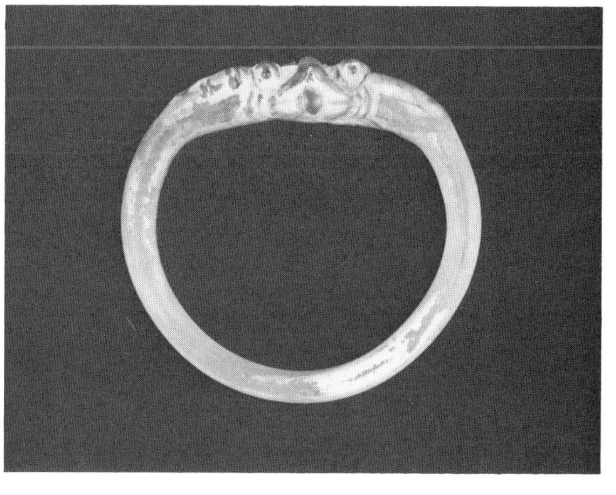

Figure 12. *Sample 5801*

have processed the barite to the carbonate or oxide, but that does not seem likely.)

Some noteworthy observations are:

1. The high-lead glasses contain two or three sample clusters which appear to differ significantly from one another.

2. The PbO and BaO do not appear to correlate either positively or negatively when the entire group of 18 samples are considered. This implies that these oxides are not associated with one another in a single batch ingredient of uniform composition and neither are they related substitutionally; that is, they were not introduced as two separate uniform materials, such that when the percentage of one increased the other would automatically have decreased. However, if one considers only the nine glasses listed just above, there is rather good positive correlation. These nine points fit the straight line:

$$(\%BaO) = 4.42 + 0.441\ (\%PbO)$$

If nos. 54 and 86 are included, the equation becomes:

$$(\%BaO) = 2.44 + 0.517\ (\%PbO)$$

3. BaO and CaO are not correlated; therefore, the barium-containing ingredient does not contain any significant amount of calcium. However, the barium does bring in a small amount of strontium.

4. The alkali, which is quite variable in the "balance" composition, does not appear to have been introduced as plant ash[8] since the MgO values are generally quite low. The MgO could have been removed if leached plant ashes were used. Otherwise, perhaps the alkali was niter.

5. The alkali does not contain Na_2O, K_2O, CaO, and MgO in the same proportions as the particular natrons we have analyzed from elsewhere[9] (for example, Egypt and East Africa) but, of course, naturally-occurring evaporites could exist in China which do match the proportions found in the glass. Analyses of *reh*, a soda evaporite used for glassmaking in India, also are low in MgO.

6. The Al_2O_3 level in the "balance" composition is consistently about 0.59% but increases when CuO or CoO are present and when the Fe_2O_3 is unusually high. The same colorants also bring in small amounts of calcium.

7. Where Fe_2O_3 is high enough to form the "black" or dark amber glasses, the CaO is also unusually high, suggesting that the colorant might have been a metallurgical slag.

8. The CaO and MgO in these glasses (excluding those with high Fe_2O_3) are closely correlated, fitting the straight line: $(CaO^* + MgO^*) = -0.287 + 0.377\ (Na_2O^* + K_2O^*)$. The * indicates the composition has been normalized to 100% after subtracting out the PbO, BaO, and colorants.

Recently FitzHugh and Zycherman reported the discovery of an early Chinese blue pigment, a copper barium tetrasilicate, $CuO.BaO.4SiO_2$.[10] This is an analogue of the well-known synthetic pigment Egyptian blue, $CuO.CaO.4SiO_2$. Following up their experiments in our laboratories, we have synthesized the Chinese blue and also a second compound which has a purple color. This new compound was found to be identical to a purple phase found intermixed with the blue pigment in the objects studied by FitzHugh and Zycherman. Our microprobe analyses indicate that the composition of the purple crystals corresponds very closely to an empirical formula $CuO.BaO.2SiO_2$, and thus the compound appears to be a disilicate. (See Fig. 20.) Preliminary experiments indicate that the blue and purple compounds form under similar, but not identical, conditions, within overlapping temperature ranges. They also have different x-ray diffraction patterns. FitzHugh and Zycherman discovered the pigments painted onto objects of the Han Period. There certainly must be significance in the fact that these Chinese pigments contain barium in place of calcium, just as do the contemporaneous Chinese glasses, when each is compared to its Western counterparts.

Figure 13. *Sample 1596*

Figure 14. *Sample 97*

Figure 15. *Left, sample 1570, right 1563*

Figure 16. *Sample 5872*

Figure 17. *Sample 5871*

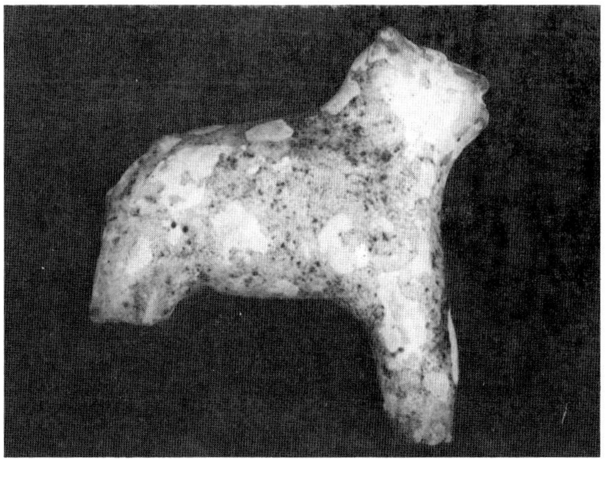

Figure 18. *Sample 1599*

Early periods: High-lead glasses without barium ($PbO:SiO_2$)

One example of a glass approximately contemporaneous with those just described was found to have a high PbO content but only trace amounts of BaO. The analysis also differs from all the other early high lead glasses in that it contains some arsenic, about 0.35% As_2O_5. The object is sample no. 82, an ear spool of the Han Period. One might expect that this object was either made outside China or made in China of glass imported from elsewhere. However, the object is a well known Chinese type, and isotope studies confirm that the lead in the glass is the same as that found in some of the leaded glasses containing BaO. Possibly it may be of a later date than that originally attributed to it, although Seligman and Beck[11] illustrate similar "capstans," said to be Han, which also do not contain barium. Examples of this type have been found in Han tombs.

A small piece of a carved bracelet, originally catalogued as heavily devitrified glass, was shown by x-ray diffraction to be made of the natural mineral tremolite, probably in the form known as nephrite. Nephrite is one of the minerals — along with jadeite, $NaAlSi_2O_6$ — which is today called *jade*. The chemical analysis (no. 96) is a close match for the theoretical composition of nephrite: $Ca_2Mg_5Si_8O_{22}(OH)_2$.

Middle Period, 7th-14th cent. A.D.: High lead ($PbO:SiO_2$); Potash-lime ($K_2O:CaO:SiO_2$); Soda-lime ($Na_2O:CaO:SiO_2$); Soda-lime-silica with lead ($Na_2O:CaO:SiO_2;PbO$)

In this group are three glasses which have very high PbO contents — more than double many of those in the preceding period — and no other major ingredient except silica. To our knowledge such compositions were not made in the West until several centuries later. Two of the glasses are small molded Buddhas, believed to be of the 8th cent. A.D., with very similar compositions, except for their colorants. The amber figure (no. 1587) is probably a "carbon amber" or, more properly, a ferri-sulfide amber.[12] The light green opaque figure (no. 1588) could possibly owe its color to the 0.20% of copper oxide (CuO) present, in combination with the tin (SnO_2) and/or antimony (Sb_2O_5), if small amounts of the yellow opacifiers $Pb_2Sb_2O_7$ or $PbSnO_3$ are present.[13] Alternatively, the opacity of this glass might be related to its arsenic content. As_2O_5 was found in only two other glasses analyzed. The glass is also unusual in that it contains traces of rubidium, Rb_2O. The third high-lead glass is the amber bead no. 65, from the Beck Collection. Its composition is very close to that of the amber Buddha, which is in the Royal Ontario Museum.

Five of the other glasses (nos. 1581, 3343, 5805, 5839, and 5844), some of which are later in date than the lead-silica glasses, have potash-lime-silica compositions. Plant ashes could possibly have been used as the alkali in making two of these glasses, however, with three of them (nos. 1581, 5805, and 5844) a question again arises about low MgO values. Potash glasses of Mediaeval dates from Western Europe invariably have MgO contents of the order of 3-5%, or even greater. Moreover, all of the dozens of analyses we have done of plant ashes themselves have correspondingly high MgO contents. If plant ashes were used for making glasses nos. 1581, 5805, and 5844, they must first have been purified by recrystallization, which removes the magnesium, calcium, and alumina from the ashes. In these glasses, some lime must then have been added back to the batch as a separate ingredient. The use of recrystallized potash and a separate lime ingredient may indicate that these objects date from the latest parts of the ranges of dates assigned to them. We usually think of alkali recrystallization as being a rather late development in European glassmaking, beginning, perhaps, not earlier than the 15th-16th centuries. It seems more likely that the Chinese glasses were made with saltpeter, which, recrys-

Figure 19. *Barium disilicate crystals in colorless, turbid glass, sample 83. Taken from JEOL Superprobe 733 electron microprobe. Similar crystals were observed and analyzed in sample 80. The scale indicates 100 microns.*

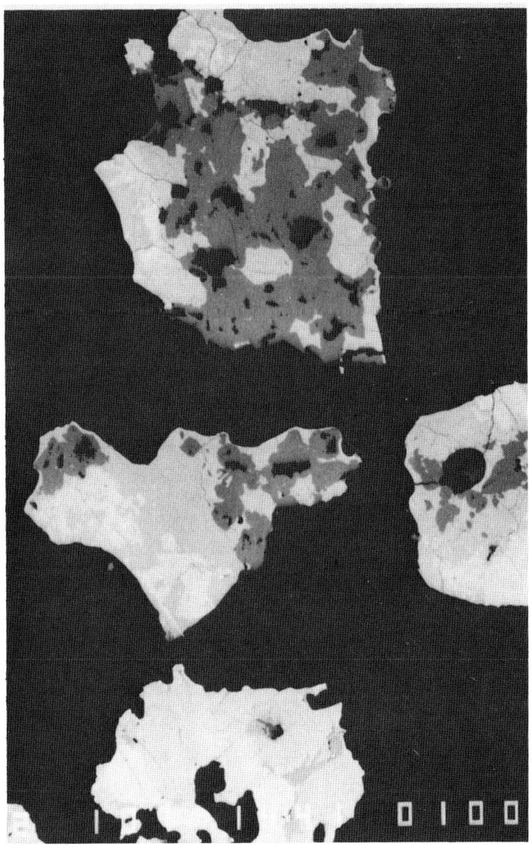

Figure 20. *Photomicrograph of a frit containing "Chinese blue" and "Chinese purple" crystals. (CMG 5792.) Crystals were grown by devitrification of a glass having the stoichiometric composition of $CuO.BaO.4SiO_2$. Devitrification was promoted by repeated crushing and heating for 48 hours at 1080°C. Dark gray patches within crystalline masses are $CuO.BaO.4SiO_2$; lighter gray patches are $CuO.BaO.2SiO_2$; black patches are SiO_2; white patches are rich in BaO and SiO_2. Compositions were determined by electron microprobe analyses by Robert Spaulding and Stephen Tong of Corning, Incorporated. Magnification approximately 80X. Similar experiments at 1160°C produced a much greater yield of the 1:1:2 purple crystals over the 1:1:4 blue crystals.*

tallized or not, might have been low in MgO. In that case, too, in order to make a stable glass, an addition of a separate lime-containing ingredient would have been required.

The analysis of no. 5839, a hairpin head, of ca. 8th century date, is more in line with earlier glasses from Western Europe because it has a higher MgO value and very high CaO. Compositionally, it puts us in mind of glasses made in Eastern Europe and Central Asia, but the object is clearly Chinese in character, and it resembles a hairpin from Boshan.[14]

The composition of another light blue opaque glass — no. 4120, a bowl in the Metropolitan Museum of Art — is quite similar to that of the hairpin. It is a potash-lime-silica glass with some lead. An x-ray diffraction pattern showed that the opacifier is calcium fluoride and the microprobe analysis established that all of the calcium is present as CaF_2, leaving an excess of flourine of only 0.03%. Very likely both elements were introduced with the same batch ingredient, namely, the mineral fluorite (fluorspar) which would have served both as a flux and an opacifier. The remaining example of a potash-lime-silica glass is no. 3343, a small, emerald-green, molded glass fish. It is certainly of Tang date, being a close parallel for a light blue transparent fish excavated in 1985 at a Tang Dynasty temple in Xian.[15] This glass contains enough MgO to have been made from potash, but it is very high in lime, unlike the Han Dynasty potash glasses analyzed by Shi Meiguang, *et al.*[16]

Sample no. 5841, from a cicada much like 5839 in its method of manufacture, is a soda-lime glass. Therefore, despite the resemblance, it represents a different chemical formulation. No. 1585 is from another hairpin head made in a similar technique, but it combines the soda and potash of the other two pieces, being a mixed alkali glass, but still having the same high CaO and MgO values. If these three pieces do have a common origin, which seems likely because of their stylistic similarity, apparently there was quite a wide variation in the types of alkali in use wherever they were made.

Sample no. 1596 is from a large, colorless bowl estimated only as somewhere between the 8th and 17th centuries. It is a soda-lime glass, but one with low CaO, and very low MgO. The K_2O value almost makes it a mixed alkali glass, but it is markedly different from those discussed above in that it contains a substantial proportion of lead. A somewhat similar composition (soda-lime-silica with lead) turns up again in a 19th century piece discussed below (no. 4108). Interestingly, the bowl no. 1596 also contains a small percentage of arsenic, which might have been used as either a decolorizer or a fining agent. If it was a deliberate addition, and not an impurity introduced with the lead, that might be taken as evidence of a rather recent date. The evidence above, along with lead isotope determinations, suggests that the bowl's date is closer to the 17th century than to the 8th century.

Late Period, 14th cent. and later:
Potash-lead-silica ($K_2O:PbO:SiO_2$);
Potash-lime-silica ($K_2O:CaO:SiO_2$);
Soda-lime-silica ($Na_2O:CaO:SiO_2$);
Soda-lime-silica with lead ($Na_2O:CaO:SiO_2:PbO$)

The twelve glasses analyzed are quite variable in composition, and all are late enough that their batch formulations might have been influenced by technological contacts with glassmakers in Europe. One piece, no. 1562, a dark blue bottle, has a soda-lime composition similar to many contemporaneous European glasses. It is also similar in that the soda value is somewhat too high and the lime value too low to form a stable glass. Consequently, this glass is badly crizzled,[18] as are many other Chinese and Western glasses with similar unstable compositions. No. 1561 is a bracelet, an example of so-called "Peking glass." It is a soda-lime glass, having a substantial addition of lead, and it seems quite stable. A small white opaque, "carved" object, no. 4110, is a soda-lime glass with a high lime content. Nos. 4112 and 4113 from a cameo vessel (dark blue over white opaque) are both potash-lime glasses with only a trace of lead. The white opaque contains 6.73% P_2O_5.

Two molded figurines (nos. 1563 and 1570) of the Buddha and of Guan Yin (the latter likely to be of 17th century date), are quite different stylistically but are quite similar in chemical composition, both being potash-lead-silica glasses. They are discussed further in another paper.[19] A third figure (no. 5871), of the Buddha, is quite different. It is much larger, done in a Southeast Asian style, and it is a bright yellowish green color. Made of a soda-lime glass colored with iron and a small amount of chromium, it also contains some arsenic. Because of the chromium, we believe it dates from the late 19th or 20th century.

An interesting, and somewhat controversial object (no. 5872), turned out to be a soda-lime glass. This is a Bodhisattva head cast in a heterogeneous, white opaque glass. Some authorities have dated it as early as the Tang or Song Dynasties, while others believe it is modern. The latter view is based on the object's unprecedented size and relative crudity of execution. It is opacified with CaF_2 and NaF, but there are precedents for this.[20] For the present, the authors are reserving judgement on the piece. The object exemplifies how little is really known about Chinese glass technology between the Tang and Ming periods, because even with an analysis in hand it is difficult to make a confident attribution.

The four objects nos. 1560, 4102, 4103, and 4104 are part of a collection of a few hundred ornamental objects purchased in China early in the 20th century. However, stylistic questions have been raised suggesting that the glasses might have been manufactured in Bohemia. Qualitative analyses of several specimens indicated that the group as a whole have widely disparate compositions. This was born out by the quantitative analyses reported here, but the compositions are all ubiquitous, and their place of origin remains unresolved. It should be noted that many are colored with uranium oxide, as is no. 1560.

Recommendations for Further Work

In conclusion, we reiterate that the analyses reported here represent a starting point, and that the limited interpretation of these findings can be expanded considerably as research in this field progresses. In that light, we invite other institutions who might have suitable Chinese glasses and related materials to contact the authors about having them analyzed.

Acknowledgements. The authors thank James Hsu and Suzanne Wilson of the Far Eastern Department of the Royal Ontario Museum for their cooperation and permission to sample objects in that collection, Leo Biek for providing the samples from the Beck Collection at Cambridge (on behalf of the Bead Study Trust), Axel von Saldern for providing the sample from Hamburg, Christine Lilyquist and Robert Koestler for providing the sample from the Metropolitan Museum of Art, as well as Robert Clague, Tran Phu, and Carol and Martin Macht for permitting us to sample their objects. We also thank Liu Yongquan and Hu Xigeng for providing ore samples. John Wosinski and Doris Evans, both of Corning Glass Works, helped with the synthesis of the Chinese blue and purple pigments. Most of the photographs were taken by Nicholas Williams and Raymond Errett of The Corning Museum of Glass. In particular, one of the authors (R.H.B.) takes this opportunity to express his gratitude to James C. Y. Watt for his encouragement and advice throughout all aspects of The Corning Museum's study of Chinese glass.

Notes

1. The term "Western World" is used here in a very loose sense, representing any location to the west of Central Asia.
2. For a survey of scientific techniques applied to the study of early glasses, see: R. H. Brill, *Proceedings of the VIIIth International Congress on Glass,* London, The Society of Glass Technology, Sheffield, England, 1968, p. 47.
3. Extracted from lecture notes distributed in China in 1982. R. H. Brill, "The Early History of Glass and Glassmaking in the Western World," in both English and Chinese, accompanied by slide notes on the scientific investigation of early Chinese glasses, and "A Chinese-English Glossary of Terms Used in the History of Glass and Glassmaking," the latter by Stephen S. Tong and James C. Y. Watt.
4. Some references in English are:
   C. G. Seligman and H. C. Beck, *Nature,* 133, 1934, p. 982; C. G. Seligman, P. D. Ritchie, and H. C. Beck, *Nature,* 138, 1936, p. 721; and C. G. Seligman and H. C. Beck, *Bulletin of the Museum of Far Eastern Antiquities,* 10 Stockholm, 1938, p. 1; K. Yamasaki, *Proceedings of the Tenth International Congress on Glass,* 9, The Ceramic Society of Japan, Kyoto, 1974, p. 15; K. Yamasaki, *Application of Science in Examination of Works of Art,* Museum of Fine Arts, Boston, 1973, p. 229; R. H. Brill, K. Yamasaki, I. L. Barnes, K. J. R. Rosman, and Migdalia Diaz, *Ars Orientalis,* XI, 1979, p. 87. There are also several excellent papers on the subject in the recent Chinese periodical literature. Some of them are: Yuan Han Chin, *Chinese Chemistry Association Annual Meeting,* 80, 1957, p. 81; Gan Fuxi; Wang, C. F.; Chin, P. W.; *Guisuanyan Xuebao, (Journal of the Institute of Ceramics),* 16, nos. 1, 2, 1978, p. 99; Yang Boda, *Wenwu (Cultural Relics),* 5, 1979, p. 76; Yang Boda, *Palace Museum Journal,* 2, 1980, p. 14; Zhen Zhuhai, *Silicate Report,* 9, no. 1, 1981, p. 96; Fan Shimin and Zhou Baozhong, *Bulletin of the Museum of Chinese History,* 5, 1983, p. 97; Zhang Fukang, Cheng Zhuhai, and Zhang Zhigang, *Journal of the Chinese Silicate Society,* 11, no. 1, 1983, p. 67.
5. See note 4 and several other chapters in this volume.
6. All the diffraction work mentioned here was done by John Geiger of Corning Glass Works. See also this volume Chapter 15.
7. See note 3.
8. For plant ash compositions see: D. Barag, R. H. Brill, A. L. Oppenheim, and A. von Saldern, *Glass and Glassmaking in Ancient Mesopotamia,* The Corning Museum of Glass, Corning, 1971, pp. 105-128.
9. R. H. Brill, *Scientific Investigations of the Jalame Glass,* in G.D. Weinberg, editor, *Excavations at Jalame, Site of a Glass Factory in Late Roman Palestine,* University of Missouri Press, 1986, Chapter 9, pp. 257-294.
10. E. W. FitzHugh and L. A. Zycherman, *Studies in Conservation,* 28, 1983, p. 15. See also this volume Chapter 5 and Addenda to Chapters 4 and 5.
11. C. G. Seligman and H. C. Beck, *op. cit.,* plate IV. See also this volume Chapter 19, in which Huang Qishan mentions similar objects made of K_2O:CaO:SiO_2 glass.
12. J. W. H. Schreurs and R. H. Brill, "Iron and Sulfur Related Colours in Ancient Glasses," *Archaeometry,* v. 26, Part 2, 1984, pp. 199-209.
13. For information on yellow opacifiers see note 2 and references cited therein. See also R. H. Brill, I. L. Barnes, and B. Adams, *Recent Advances in Science and Technology of Materials,* v. 3, Plenum Press, New York and London, 1974, p. 9.
14. This volume, Chapter 7.
15. Private communication, An Jiayao.
16. Shi Meiguang, He Ouli, and Zhou Fuzheng, "Investigation of Some Chinese Potash Glasses Excavated from Tombs of the Han Dynasty," *Journal of the Chinese Silicate Society,* v. 14, no. 3, 1986, pp. 307-313.
17. This volume, Chapter 5.
18. R. H. Brill, *Conservation in Archaeology and the Applied Arts: Preprints of the Contributions to the Stockholm Congress 2-6 June 1975,* The International Institute for Conservation of Historic and Artistic Works, 1975, p. 121; R. H. Brill, *Journal of Glass Studies,* v. 20, 1978, pp. 100-118.
19. See note 17.
20. For example, see sample no. 2076, this volume, Chapter 5.

CATALOGUE OF GLASS SAMPLES

Studies completed:
- * quantitative analysis (aa)
- e quantitative analysis (microprobe)
- s qualitative analysis (spectrographic)
- \+ Pb isotope analysis
- p some physical properties

Early Periods

* s +	05	*Bi* disk, 4th-3rd cent. B.C. Colorless glass, heavily weathered. CMG 51.6.567. (3330, Pb-1005)
* s e +	06	Weathering product from 05. (3331, Pb-1006)
* +	07	*Bi* disk, 4th-3rd cent. B.C. (?). Turbid green glass, lightly weathered. CMG 51.6.548. (Pb-1007)
s +	09	Sword scabbard terminal, 2nd cent. B.C. Slightly turbid colorless glass, heavily weathered. CMG 62.6.15. Emission spectrography shows ±5% PbO; ±.4% BaO. (3329, Pb-1009)
* s	48	Flattened bar, uncertain date. White opal glass with red swirls, moderately weathered, 7.3 x 2.2 x 0.9 cm. White color appears to have been struck. CMG study cards 1313 a & b. White glass. (3332)
	49	Red opaque glass from same object as 48. (3333)
* s +	50	Eye bead fragment, 4th-3rd cent. B.C., possibly from Jincun near Luoyang. Dark blue transparent glass with white opaque decoration marvered in, moderately weathered. CMG 51.6.549B. (3334, Pb-1550)
* s +	51	Large bead, Han (?). Light blue transparent glass apparently undecorated, heavily weathered. CMG 67.6.2. (3342, Pb-1551)
* s +	52	Eye bead, 4th-3rd cent. B.C. Dark blue transparent glass with yellow opaque and white opaque decoration marvered in, very lightly weathered. CMG 672. (3335, Pb-1552)
+	53	Yellow opaque glass from same object as 52. (Pb-1553)
* s +	54	Eye bead, 4th-3rd cent. B.C. Dark blue transparent glass with inlaid white (?) opaque and yellow (?) opaque decoration, moderately weathered. CMG 671. (3336, Pb-1554)
+	57	Glass fish, flat with incised decoration, probably Han. White opaque glass, moderately weathered. CMG 51.7.3. (3338, Pb-1557)
+	59	Glass cicada, Han. Greenish glass, heavily weathered. Submitted by Rutherford J. Gettens, Freer Gallery of Art SC-543. (240, Pb-1559)
s +	64	Glass cicada, Han. Turbid white glass, heavily weathered. CMG 51.6.558. (3341, Pb-1564)
s	66	Eye bead, Han (?). Light blue turbid glaze over frit-like body material, moderately weathered. Beck Collection "MS 706 B/2454." (5862)
s +	68	Eye bead, Han (?). Light blue transparent glass, moderately weathered. Beck Collection "C95317, sp. gr. 3.27." (5867, Pb-1568)

*	s		+		69	Eye bead, Han (?). "Black," frit-like body with light blue turbid inlay eye and remains of white fill. Beck Collection "CGS, E, B/2455, sp. gr. 3.1." (5864, Pb-1569)
		e	+		73	Crescent with eye decoration, 4th-lst cent. B.C. Dark glass, heavily weathered. ROM 933.12.167. (5808, Pb-1573)
*	s				74	Glass plaque with running animal, 2nd-1st cent. B.C. (?) Turbid pale green glass, heavily weathered. ROM 930.21.215. (5833)
		e		p	75	Cylindrical bead, date uncertain. Somewhat turbid colorless glass, lightly weathered. ROM 960.241.231. (5813)
			+		76	Eye bead, 5th-1st cent. B.C. Blue (?) glass with decoration marvered in, heavily weathered. ROM 933 x 84.1. Sample is of yellow opaque glass. (5816, Pb-1576)
			+		77	Glazed bead, 4th-1st cent. B.C. Frit-like body with weathered green glaze, "S" and circle decoration. ROM 933.84.11. (5826, Pb-1577)
			+		78	Eye bead, 4th-1st cent. B.C. "Black" glass with inlay eyes and aqua glass remaining in some places and double bands of dotted lines once containing a white fill; moderately-heavily weathered. Sample is of black glass. ROM 933.84.7. (5818, Pb-1578)
*	s				79	Aqua glass from same object as 78. (5819, Pb-1579)
*	s	e	+	p	80	*Bi* disk, broken and repaired, 3rd cent. B.C. Slightly turbid (?) pale green glass, heavily weathered. ROM 971.144.4. (5830, Pb-1580)
*	s	e	+	p	82	Ear spool, Han. "Black" glass, heavily weathered. ROM 933.84.13. (5832, Pb-1582)
*	s	e	+	p	83	Glass plaque with dragon, 2nd-1st cent. B.C. Turbid colorless glass, heavily weathered. ROM 933.23.340. (5835, Pb-1583)
*	s	e	+	p	84	Glass cicada, Han. Turbid pale greenish glass, heavily weathered. ROM 930 x 170.9. (5837, Pb-1584)
*	s		+		86	Vessel fragment, 4th-1st cent. B.C. Light blue opaque glass with applied dots of same color set over white and dark frit liners, moderately weathered. Underside may contain traces of core residue. One of six similar fragments. ROM 933.58.5 (5842, Pb-1586). Note: A beaker-shaped vessel, made up of fragments which one of the authors (R.H.B.) believes came from the same ancient object as this sample, is now in the hands of Eskenazi, Ltd. (Spring, 1990).
*	s		+		89	Barrel-shaped bead, 4th-1st cent. B.C. Reddish-brown glass body, (original color indeterminate), with applied eyes of light blue translucent glass, heavily weathered. ROM 933.12.165. (5848, Pb-1589)
	s		+		90	Eye bead, 4th-1st cent. B.C. Friable gray body of pumice-like appearance; glazed surface with dotted decoration of indeterminate original color, heavily weathered. Sample is of body material. Emission spectrography showed glaze to contain some lead. ROM 960.241.248. (5851, Pb-1590)
*	s	e	+	p	91	Eye bead, 5th-1st cent. B.C. Medium blue glass body, inlay eye decoration missing, heavily weathered. ROM 933.84.3. (5852, Pb-1591)
*	s		+		92	Bead, 5th-1st cent. B.C. Greenish-blue glass, heavily weathered. Might once have contained eye decoration or might have been faceted. ROM 933.84.4. (5854, Pb-1592)

* s + p	93	Small bead, 4th-1st cent. B.C. Black glass, probably once contained eye decoration; heavily weathered. ROM 933 x 84.5. (5855, Pb-1593)
* s +	94	Eye bead, 4th-1st cent. B.C. Friable terra cotta (?) body, black glazed surface, traces of green fill (probably glass) and white liner remain, also dotted decorations of white fill, heavily weathered. Sample is of black glaze. ROM 933 x 84.6. (5856, Pb-1594)
* s +	95	Cube-shaped bead with raised eye decorations at each corner, 4th-1st cent. B.C. Colorless glass with blue glass eyes; heavily weathered. ROM 933 x 84.8. (5858, Pb-1595)

Middle Periods

* s	65	Small bead, Tang. Amber or olive glass, unweathered. Original diameter 4mm. Beck Collection "1-1938-7." (5861)
* s + p	1581	Glass hairpin, 15th-16th cent. A.D. Light blue opaque glass, lightly iridescent in places. ROM 960.241.230. (5831, Pb-1581)
* s e + p	1585	Hairpin head, molded relief design, 7th-9th cent. A.D. Somewhat turbid colorless glass, moderately weathered. ROM 930.170.3. (5840, Pb-1585)
* s + p	1587	Small glass Buddha, 7th-8th cent. A.D. Amber glass, moderately weathered. ROM 925.26.367A. (5845, Pb-1587)
* s + p	1588	Small glass Buddha, 7th-8th cent. A.D. Light green opaque glass, moderately weathered. ROM 925.26.367L. (5846, Pb-1588)
* s + p	1596	Large bowl, Tang-Ming or 8th-17th cent. A.D. Said to be from tomb in temple ground at Luoyang. Colorless transparent glass with brownish substance and rough grayish white deposit on parts of interior. Diameter 21.3 cm. ROM 923.24.163. (5860, Pb-1596)
* p	3343	Glass fish, Tang. Emerald green glass, moderately weathered. CMG 51.6.550. Density = 2.563 g/cc.
e + p	4120	Glass bowl, prob. 13th-14th cent. Light blue opaque glass, moderately weathered. Acquired in Cairo. Metropolitan Museum of Art 26.7.1273. Illustrated in Addendum to Chapter 5. (Pb-2020).
p	5801	Glass bracelet, 7th-9th cent. A.D. Colorless glass, lightly weathered. ROM 921.21.355A. (5801)
p	5803	"Carved" plaque, 2nd-8th cent. A.D. Amber glass, lightly weathered. ROM 928.12.135. (5803)
e p	5805	Cicada pendant, 8th-12th cent. A.D. Colorless transparent glass, lightly weathered. ROM 930.170.5. (5805)
p	5811	Glass peach, 10th-14th cent. A.D. Very pale greenish, slightly turbid glass, moderately weathered. ROM 933 x 84.12. (5811)
* s e p	5839	Hairpin head, 7th-9th cent. A.D. White opaque glass, lightly weathered. ROM 960.241.233. (5839)
* s e p	5841	Cicada fragment, molded, 7th-9th cent. A.D. Turbid colorless glass, lightly weathered. ROM 960.241.247. (5841)
* s p	5844	Wound and lobed fragment of uncertain use and date. Light blue transparent glass with white opaque swirls, moderately weathered. ROM 918.21.913. (5844)

*	p 5872	Glass head of a Bodhisattva. Date uncertain; various scholars have dated it from as early as the Tang/Song Dynasties until modern times. Cast, possibly from a stone prototype, in heterogeneous white opaque glass, with many voids, and three types of surface accretion. Opacified with CaF_2 and NaF. Overall height ~20 cm. From the collection of Drs. Carol and Martin Macht, presently on loan to the Cincinnati Art Museum.

Late Periods

* s	+ p 1560	Large glass bead, probably early 20th cent., acquired in China. Possibly made in China but more likely imported from Europe. Bright green transparent glass colored with uranium, unweathered. Diam. 2.8 cm, drilled bore. CMG 51.6.246. (4100, Pb-1560)
* s	+ 1561	"Peking glass" bracelet, faceted, 19th or early 20th cent. Dark blue transparent glass, incipient crizzling (?) CMG unaccessioned. (4108, Pb-1561)
* s	+ 1562	Bottle, 18th-19th cent. Dark blue transparent glass, crizzled with surface spalling away. Submitted by Axel von Saldern, Museum für Kunst und Gewerbe, Hamburg. (4111, Pb-1562)
* s e	+ p 1563	Glass Buddha, mold cast with engraved details, 17th-19th cent., probably Thailand. Gray transparent glass with loose, brownish accretion. See A.B. Griswold, "A Glass Image and Its Associations," *Journal of Glass Studies*, vol. 5, 1963, pp. 75-103. CMG 56.6.10. (5870, Pb-1563)
e	+ p 1570	Guan Yin, Song-Ming, or 12th-17th cent. Slightly turbid colorless glass, some evidence of light weathering or crizzling. ROM 922.20.80A. (5802, Pb-1570)
* s	p 4102	Small carved bead, probably early 20th cent. Acquired with 4100 (1560 above). Colorless glass. Diam. 1.2 cm, drilled bore. CMG 51.6.250 (?). (4102)
* s	p 4103	Small carved bead, probably early 20th cent. Acquired with 4100 (1560 above). Purple transparent glass. Diam. *ca.* 1.1 cm, drilled bore. CMG no. illegible. (4103)
* s	p 4104	Small uncarved bead, probably early 20th cent. Acquired with 4100 (1560 above). Pale blue transparent glass. Diam. 1.1 cm, drilled bore. CMG 51.6.267. (4104)
* s	p 4110	Small fragment of carved glass, probably 18th-19th cent. (?) White opaque glass, incipient crizzling. CMG unaccessioned. (4110)
* s	+ 4112	Cameo-carved vessel, 19th cent. White opaque base glass. Submitted by Robert H. Clague. (Pb-1567)
* s	+ 4113	Same object as 4112. Dark blue transparent overlay glass. (Pb-1566)
*	p 5871	Large glass Buddha, probably 19th cent. Green (slightly yellowish) glass, with traces of gilding, little or no weathering. 32 cm tall. In the collection of Mr. Tran Phu, who acquired the object in Thailand. CMG loan L137.6.85. Density = 2.470 g/cc; ref. index = 1.507.

Non-Glasses and Non-Chinese

*		+	08	Fragment of glazed ceramic, 18th cent., made at Jingdezhen. Sample is of bright yellow glaze. Collection of E.T. Hall, Oxford. (Pb-1008)
*		+	56	Large eye bead, probably Near Eastern 600-250 B.C. Turbid light blue glass with large knobbed black and white eyes and numerous other applied blobs, heavily weathered. Sample is of weathered yellow opaque glass knob. CMG unaccessioned. (3337, Pb-1556)
		+	72	Metal buckle (probably bronze), gilded, with blue opaque glass inlays, 8-17th cent. A.D. Sample is of black metallic corrosion products. ROM 950.77.3. (5807, Pb-1572)
*	s	p	96	Curved rod with spiralling, possibly from a bracelet, 6th-4th cent. B.C. (?) or later. Catalogued as heavily devitrified or weathered glass. Now known to be made of nephrite. ROM 933 x 84.14. (5847)
	e	+	97	Faceted bar of "Chinese purple" (CuO.BaO.4SiO), Han (?). Sintered molded bar of lavender color, slightly weathered. Electron microprobe analysis shows presence of some lead. ROM 981.5.6. (5799, Pb-1597) Note: See also samples Pb-2000, 2001, 2002, and 2003 in companion paper. These are other examples of Chinese purple.
		+	1599	Glazed terra cotta figure of an animal, date uncertain, possibly Chinese; purchased in Ghazni. Sample is of weathered green glaze. R.H.B. collection. (Pb-1599)

Ores and Minerals

		+	1804	Bournonite (PbCuSbS). Guangdon Province. Courtesy Paul E. Desautels, National Museum of Natural History, Washington; NMNH 90614. (Pb-1804)
		+	1805	Galena, Shuikoushan, Hunan Province, Courtesy John E. Dayton. (Pb-1805)
		+	1806	Galena, Jiangxi Province. Courtesy Liu Yongquan, Geological Museum, Beijing. (Pb-1806)
		+	1807	Galena, Liaoning Province. Courtesy Liu Yongquan. (Pb-1807)
		+	1808	Galena, Yunnan Province. Courtesy Liu Yongquan. (Pb-1808)
		+	1809	Galena, Hunan Province. Courtesy Liu Yongquan. (Pb-1809)
	s		1598	Remingtonite (?) (hydrated cobalt carbonate), prob. Yimen Mine in Yunnan or Jinchuan Mine in Gansu. Courtesy Hu Xigeng, Central-South Institute of Mining and Metallurgy, Changsha. Emission spectrographic analysis shows ~2% Pb, ~1% Ba and major Mn, Fe, and Al. (4780, Pb-1598)

APPENDIX: ANALYTICAL METHODS

BRANDT A. RISING, Umpire and Control
STEPHEN S.C. TONG, Corning, Inc.

1. Preparation of samples for analysis:
   Chips and pieces of glass received for analysis are ground to a fine powder in a clean agate mortar.

2. Spectrographic analysis:
   a. Approximately 70 mgs of the ground sample is mixed with graphite (2 parts graphite: 1 part sample) and four graphite spectrographic electrodes are loaded with the mix.
   b. Spectrographic analysis of the samples is conducted using a DC arc, 10 AMPS current and burning to completion (approximately 6 minutes). The resulting spectra are recorded on spectrographic plates which are kept on permanent file.
   c. Calibration of the spectrographic analysis is obtained using at least two sets of synthetic glass standards with assigned values for all trace elements normally reported.

3. Quantitative analysis of major glass elements:
   a. All elements of significant concentration which exceed the working range of the spectrographic glass standards are determined by atomic absorption analysis versus specially prepared standards.
   b. For atomic absorption analysis, 100 mgs of fine ground sample is weighed out and decomposed in a plastic beaker with a mixture of HC1 and HF. Following volatilization of the silica, the samples are repeatedly taken to dryness on the water bath to remove excess HF. The dried salts are dissolved in hydrochloric acid solution and diluted appropriately for subsequent determination of the various elements.
   c. Calibration of the atomic absorption instrument is obtained using synthetic glass standards and by vigorous control of solution matrices. In addition to the standard glasses carried through each analysis, selected reference glasses are also included to provide a cross check on possible biases which may affect the long term comparability of the results.

4. Electron microprobe analysis:
   Small chips of glass specimens were mounted in epoxy resin, polished to 0.25 nm finish and analyzed with an Applied Research Laboratory SEMQ electron microprobe. Glasses containing appropriate amounts of various elements were used as standards. In most cases, a 70 nm diameter electron beam was used for analysis. However, a much smaller beam diameter was used for some samples such as sample number 83.

Charles Nielson of JEOL, USA assisted in some of the microprobe analyses.

Table 1

Analyses of Chinese Glasses (Early Periods)

	2-1 BC plaque clrls. *83*	2BC-2AD plaque pale grn. *84*	3 BC *bi* clrls. *80*	bead dk. bl. *52*	4-1 BC crescent dark (?) *73*	5-1 BC bead grn.bl. 92	4-1 BC bead blk. 93
SiO_2	~40.5	~42.4	~36.8	~42.0	~40.3	~41.4	~37.3
Na_2O	2.72	3.55	1.87	1.58	3.17	5.94	3.75
CaO	0.96	0.40	0.46	1.92	3.67	1.37	1.89
K_2O	0.22	0.13	0.16	0.17	0.57	0.16	0.37
MgO	0.035	0.04	0.15	0.18	0.61	0.58	0.61
Al_2O_3	0.18	0.23	0.28	2.12	1.82	0.89	1.19
Fe_2O_3	0.24	0.32	0.14	0.63	0.65	0.27	7.35
TiO_2	0.005	0.01	nf	0.08		0.01	0.05
Sb_2O_5	nf	0.01	nf	0.06		0.05	0.01
MnO	0.003	0.004	0.003	0.003		0.004	0.01
CuO	0.01	0.05	0.02	0.10	0.13	2.07	0.42
CoO	nf	nf	nf	0.02		nf	nf
SnO_2	nf	nf	nf	0.10		0.003	0.05
Ag_2O	0.03	0.05	0.02	0.08		0.01	0.01
PbO	35.2	33.3	42.6	31.6	32.2	37.4	37.5
BaO	19.7	19.2	17.4	19.2	17.4	9.71	9.40
SrO	0.20	0.20	0.10	0.10	0.15	0.1	0.05
Li_2O	0.001	nf	0.005	nf		nf	0.01
Rb_2	nf	nf	nf	nf		nf	nf
B_2O_3	nf	nf	nf	<0.01		nf	nf
V_2O_5	nf	nf	nf	nf		nf	nf
Cr_2O_3	nf	nf	nf	<0.005		nf	nf
NiO	nf	nf	nf	nf		nf	nf
ZnO	0.008	0.12	0.007	0.005		0.049	0.022

Table 1

Analyses of Chinese Glasses (Early Periods)

	2-1 BC	2BC-2AD	3 BC		4-1 BC	5-1 BC	4-1 BC
	plaque	plaque	*bi*	bead	crescent	bead	bead
	clrls.	pale grn.	clrls.	dk. bl.	dark (?)	grn.bl.	blk.
	83	*84*	*80*	*52*	*73*	92	93
ZrO_2	nf	nf	nf	<0.01		nf	nf
Bi_2O_3	nf	0.005	nf	0.005		0.005	0.01
As_2O_5	nf	nf	nf	nf		nf	nf
	m, aa	m, aa	m, aa		m		

Keys to Tables: aa = atomic absorption; m = microprobe; s = spect.; ps = partial spect.; unmarked = aa + s.
c = crizzled; ic = incipient crizz.; min = mineral; xrd = x-ray diff.; u = minor U_2O_3;
f = Ca as CaF_2; P = 6.73% P_2O_5; P' = 1.45% P_2O_5.

	4-1 BC	4-1 BC					
	bead	bead	bead	bar	bead	cicada	bead
	blk.	blk. (?)	clrls.	wt. op.	aqua	wt. op.	lt. bl.
	94	89	*75*	*48*	*79*	*64*	66
SiO_2	~41.7	~51.5	~69.5	~53.0	~53.0		
Na_2O	2.02	5.39	4.26	3.26	5.81	0.08	
CaO	2.92	1.00	7.31	27.7	0.79	0.5	
K_2O	0.63	0.21	16.6	1.95	0.13	nf	
MgO	0.53	0.75	0.38	7.44	0.55		
Al_2O_3	1.90	0.98	1.06	5.50	0.43		
Fe_2O_3	5.04	3.54	0.31	0.98	0.45		
TiO_2	0.10	0.03		0.15			
Sb_2O_5	0.01	nf		nf			

MnO	0.036	0.008		0.01	~0.008		
CuO	0.35	0.60		0.03			
CoO	0.05	nf		nf			
SnO_2	0.02	0.05		nf			
Ag_2O	0.02	0.01		0.003			
PbO	34.5	21.0	0.02	0.01	23.0	major	~10
BaO	10.1	14.5	0.08	<0.01	15.8	minor	~13
SrO	0.1	0.1		<0.01		nf	
Li_2O	0.005	nf		<0.005		nf	
Rb_2O	nf	nf		<0.01		nf	
B_2O_3	nf	nf		<0.01			
V_2O_5	nf	nf		<0.005			
Cr_2O_3	nf	nf		<0.005			
NiO	nf	nf		nf			
ZnO	0.005	0.010			0.014		
ZrO_2	nf	nf		<0.01			
Bi_2O_3	nf	0.30		nf			
As_2O_5	nf	nf		nf			
			m		aa, ps	ps	ps
			5-4 BC	2-1 BC	2-1 BC	4-1 BC	5-1 BC
	bead	bead	rod	w.prod.	plaque	bead	bead
	lt. bl.	(body)	(devit.)	from 05	not glass	crls.	med. bl.
	68	*90*	96	*06*	*74*	95	*91*
SiO_2			~58.5	~42.8	-0-	~51.3	~52.4
Na_2O		0.X	0.12	0.46	0.77	6.12	10.1
CaO		major	12.8	2.11	35.2	0.37	1.48
K_2O		0.X	0.06	0.043	0.05	0.084	0.16
MgO		0.X	25.7	0.062	0.52	1.52	2.62
Al_2O_3		0.X	1.28	0.90	0.06	0.46	1.21
Fe_2O_3		0.X	1.31	0.10	0.08	0.10	0.28

			5-4 BC	2-1 BC	2-1 BC	4-1 BC	5-1 BC
	bead	bead	rod	w.prod.	plaque	bead	bead
	lt. bl.	(body)	(devit.)	from 05	not glass	crls.	med. bl.
	68	*90*	96	*06*	*74*	95	*91*
TiO_2		0.10	0.10	<0.01	nf	0.01	0.03
Sb_2O_5		nf	nf	nf	nf	0.05	0.01
MnO		0.01	0.087	0.001	0.002	0.002	0.004
CuO		0.05	0.005	0.02	0.005	0.01	1.31
CoO		nf	nf	nf	nf	nf	nf
SnO_2		0.005	nf	nf	nf	0.003	0.005
Ag_2O		0.005	<0.001	0.005	nf	0.001	<0.005
PbO	~20	Minor	0.01	48.7	0.005	28.3	19.2
BaO	~7	0.15	0.05	4.68	0.01	11.4	11.1
SrO		nf	nf	0.10	0.01	0.3	0.1
Li_2O		nf	nf	nf	nf	nf	nf
Rb_2O		nf	nf	nf	nf	nf	nf
B_2O_3		0.01	nf	<0.01	nf	nf	
V_2O_5		nf	0.005	<0.005	nf	nf	nf
Cr_2O_3		nf	nf	<0.005	nf	nf	nf
NiO		nf	nf	nf	nf	nf	nf
ZnO			0.017	0.004	0.013	0.002	0.016
ZrO_2		nf	nf	nf	nf	nf	nf
Bi_2O_3		nf	nf	nf	nf	0.002	nf
As_2O_5		nf	nf	nf	nf	nf	nf
	ps	s			min		m, aa

		4-1 BC					2 BC-2 AD
	bi	vessel	bead	bead	bead	bead	spool
	clrls.	lt.bl.op.	dk. bl.	dk. bl.	lt. bl.	blk.	blk.
	05	*86*	*50*	*54*	*51*	*69*	*82*
SiO_2	~67.7	~60.7	~55.0	~66.2	~37.2	~46.9	~27.9
Na_2O	1.20	1.63	7.53	10.6	3.94	3.01	0.22
CaO	0.40	2.45	2.95	4.50	1.01	2.67	0.88
K_2O	0.13	4.57	4.05	7.37	0.02	0.66	0.16
MgO	0.062	0.68	1.27	2.09	1.17	0.73	0.01
Al_2O_3	0.42	4.28	2.13	2.29	0.15	1.93	0.26
Fe_2O_3	0.10	0.53	0.62	1.26	0.01	4.63	8.21
TiO_2	<0.01	0.15	0.08	0.10	<0.01	0.12	0.1
Sb_2O_5	nf	nf	0.06	0.06	0.01	0.05	0.01
MnO	0.005	0.035	0.01	0.03	<0.005	0.02	0.005
CuO	0.03	1.31	0.31	0.42	0.30	0.50	0.06
CoO	nf	nf	0.02	0.04	nf	0.08	nf
SnO_2	nf	0.15	1.00	0.70	0.10	0.20	0.003
Ag_2O	0.15	0.02	0.05	0.01	0.001	0.02	0.01
PbO	16.5	12.6	15.0	3.41	41.4	35.2	61.9
BaO	13.2	18.4	9.74	0.70	14.6	3.09	0.01
SrO	0.10	0.1	0.10	0.10	0.10	0.05	0.01
Li_2O	nf	nf	0.01	0.05	nf	<0.001	nf
Rb_2O	nf	nf	nf	0.05	nf	nf	nf
B_2O_3	<0.01	nf	<0.01	0.01	<0.01	0.01	nf
V_2O_5	nf	0.005	nf	<0.005	nf	nf	nf
Cr_2O_3	<0.005	nf	<0.005	<0.005	nf	nf	nf
NiO	<0.005	nf	<0.005	<0.005	nf	nf	0.005
ZnO	0.004	0.007	0.004	0.006	0.011	0.022	0.02
ZrO_2	nf	nf	<0.01	<0.01	nf	<0.01	nf
Bi_2O_3	nf	nf	0.001	nf	0.001	0.10	nf
As_2O_5	nf	nf	nf	nf	0.01	nf	0.35
							m, aa

Table 2

Analyses of Chinese Glasses (Middle Periods; Tang-Yuan)

	7-8 c.	7-8 c.	Tang bead	7-9 c.	7-9 c.	Tang (?)
	amber	lt.gr.op.	amber	clrs.	crls.	clrls.
	1587	*1588*	65	*5841*	*1585*	*1596*
SiO_2	~19.6	~21.3	~29.0	~62.8	~65.1	~70.1
Na_2O	0.20	0.21	0.33	9.20	8.84	9.81
CaO	1.00	0.28	0.25	16.9	10.6	5.68
K_2O	0.01	0.01	0.01	1.15	9.76	3.32
MgO	0.32	0.11	0.11	4.12	2.71	0.23
Al_2O_3	1.03	0.17	0.47	4.91	1.96	0.82
Fe_2O_3	2.52	0.16	2.14	0.41	0.56	0.27
TiO_2	0.03	nf	0.05	0.30	0.05	0.10
Sb_2O_5	0.01	0.50	nf	nf	nf	nf
MnO	0.096	0.003	0.02	0.009	0.024	0.12
CuO	0.12	0.20	0.15	0.03	0.01	0.01
CoO	nf	nf	nf	nf	nf	nf
SnO_2	0.005	0.90	0.001	0.003	nf	nf
Ag_2O	0.04	0.03	0.001	<0.001	<0.005	0.001
PbO	75.0	75.9	67.4	0.01	0.37	9.06
BaO	nf	0.06	<0.01	0.05	0.01	0.20
SrO	nf	nf	nf	0.1	nf	0.04
Li_2O	nf	nf	nf	0.003	0.01	0.001
Rb_2O	nf	0.10	nf	nf	nf	nf
B_2O_3	nf	nf	0.01	nf	nf	0.02
V_2O_5	nf	nf	nf	0.005	nf	nf
Cr_2O_3	nf	nf	nf	nf	nf	nf
NiO	nf	nf	nf	nf	nf	nf
ZnO	0.009	0.005	0.034	0.038	0.024	0.023
ZrO_2	nf	nf	<0.01	0.01	nf	nf
Bi_2O_3	0.005	0.005	0.001	nf	nf	0.005
As_2O_5	nf	0.10	nf	nf	nf	0.15
				m, aa	m, aa	

	7-9 c. wt. op. *5839*	7-9 c. em. grn. *3343*	8-12 c. clrls. *5805*	7-17 c. lt. bl. *5844*	15-16 c. lt.bl.op. *1581*	14-15 c. lt.bl.op. *4120*	7-13 c.(?) wt. op. *5872*
SiO_2	~64.4	~62.0	~68.1	~67.4	~68.7	~56.7	~68.6
Na_2O	0.99	1.24	0.73	0.72	0.05	0.40	17.1
CaO	18.7	15.5	8.75	8.36	7.71	8.78	7.5
K_2O	8.07	13.6	18.6	20.2	15.6	16.2	1.46
MgO	5.49	4.33	1.29	0.31	0.17	0.40	0.24
Al_2O_3	1.51	1.25	0.78	1.49	0.70	0.66	4.46
Fe_2O_3	0.50	0.95	0.60	0.42	0.23	0.90	0.28
TiO_2	0.05	0.10		0.03	0.03		0.03
Sb_2O_5	nf	nf		nf	nf	0.27	nf
MnO	0.008	0.014		0.009	0.002	0.05	0.11
CuO	0.005	0.55		0.56	1.53	1.24	0.01
CoO	nf	nf		nf	nf		nf
SnO_2	nf	0.08		0.005	nf	0.51	nf
Ag_2O	<0.005	nf		<0.005	0.04		nf
PbO	0.03	0.05	0.03	0.03	5.21	13.9	0.08
BaO	0.10	0.01	0.09	0.01	0.01		0.10
SrO	0.01	0.02	0.02	nf			0.01
Li_2O	0.01	0.001		nf	<0.001		0.005
Rb_2O	nf	nf		nf	nf		nf
B_2O_3	nf	<0.01		0.01	nf		0.02
V_2O_5	nf	0.005		nf	nf		<0.005
Cr_2O_3	nf	nf		nf	nf		nf
NiO	nf	nf		nf	nf		nf
ZnO	0.10	0.30		0.41	0.02		0.017
ZrO_2	nf	<0.01		nf	nf		<0.01
Bi_2O_3	nf	nf		nf	0.005		nf
As_2O_5	nf	nf		nf	nf		nf
	m, aa		m			m,f	xrd

Note: For no. 1581 and other opaque glasses, calcium is reported as CaO, but could possibly be present largely as CaF_2.

Table 3

Analyses of Chinese Glasses (Late Periods; Ming and Following)

	12-17 c. clrls. *1570*	17-19 c. grayish t. *1563*	19 c. green *5871*	18-19 c.(?) wt. op. *4110*	18-19 c. dk. bl. *1562*
SiO_2	~44.2	~40.0	~71.5	~58.5	~73.6
Na_2O	0.22	0.25	16.0	2.56	19.1
CaO	0.34	0.27	9.28	21.1	4.52
K_2O	16.6	10.6	0.26	10.1	0.41
MgO	0.05	0.12	0.12	5.25	0.24
Al_2O_3	0.31	0.17	0.34	1.72	1.55
Fe_2O_3	2.60	0.15	1.47	0.51	0.24
TiO_2		nf	0.05	0.10	0.10
Sb_2O_5		nf	nf	nf	nf
MnO		0.025	0.69	0.012	0.005
CuO		0.01	0.01	0.005	0.005
CoO		nf	nf	nf	0.15
SnO_2		nf	nf	nf	nf
Ag_2O		0.01	<0.001	<0.001	<0.001
PbO	36.0	48.5	0.01	0.04	0.04
BaO		0.10	0.01	0.07	0.002
SrO	0.12	nf	nf	0.03	0.02
Li_2O		nf	nf	0.001	<0.001
Rb_2O		nf	nf	nf	nf
B_2O_3		nf	0.02	nf	nf
V_2O_5		nf	nf	nf	nf
Cr_2O_3		nf	0.03	nf	nf
NiO		nf	nf	0.005	0.005
ZnO		0.010	0.033	0.009	0.012
ZrO_2		nf	nf	<0.01	<0.01
Bi_2O_3		0.03	nf	nf	nf
As_2O_5		nf	0.15	nf	nf
	m	m, aa		ic	c

	19-20 c. dk. bl. *1561*	19-20 c. wt. op. *4112*	19-20 c. dk. bl. *4113*	20 c.(?) ylw-grn. *1560*	20 c.(?) clrls. *4102*	20 c.(?) purple *4103*	20 c.(?) pale bl. *4104*
SiO_2	~57.3	~65.3	~69.9	~62.1	~42.3	~70.8	~71.0
Na_2O	13.9	3.54	3.54	2.38	0.68	1.70	17.4
CaO	6.52	8.49	7.83	1.86	0.76	6.39	5.65
K_2O	1.27	13.24	14.33	14.9	10.9	18.5	0.76
MgO	0.46	0.42	0.41	0.093	0.074	0.17	0.26
Al_2O_3	1.61	0.52	0.44	0.26	0.13	1.11	4.31
Fe_2O_3	0.24	0.26	0.24	0.17	0.12	0.26	0.19
TiO_2	0.05	0.002	0.002	0.02	0.001	0.08	0.20
Sb_2O_5	0.01	nf	nf	nf	nf	nf	nf
MnO	0.082	0.15	0.11	0.0088	0.034	0.57	0.0088
CuO	0.63	0.005	0.003	0.15	0.01	0.005	0.005
CoO	0.01	nf	0.25	nf	nf	0.01	0.01
SnO_2	0.03	0.50	0.35	nf	nf	nf	nf
Ag_2O	<0.01	0.001	0.003	0.01	0.03	0.01	0.01
PbO	17.5	0.42	0.54	17.8	44.9	0.08	0.02
BaO	0.10	0.01	0.01	0.10	nf	0.10	0.05
SrO	0.005	0.01	0.01	<0.005	nf	0.01	0.02
Li_2O	0.001	0.001	0.001	0.003	0.003	0.08	0.01
Rb_2O	nf	0.01	0.01	nf	nf	nf	nf
B_2O_3	0.03	0.01	0.01	0.02	nf	nf	nf
V_2O_5	0.005	nf	nf	0.005	nf	<0.005	<0.005
Cr_2O_3	0.005	nf	nf	nf	nf	nf	nf
NiO	0.005	nf	0.05	nf	nf	0.01	nf
ZnO	0.27	0.012	0.008				
ZrO_2	<0.01	nf	nf	0.15	0.07	0.07	0.10
Bi_2O_3	nf	nf	nf	nf	nf	nf	nf
As_2O_5	nf	0.40	0.50				
P_2O_5		6.73	1.45				
	ic	p	p	u			

Table 4. Compositions of Chinese glasses in various historical periods

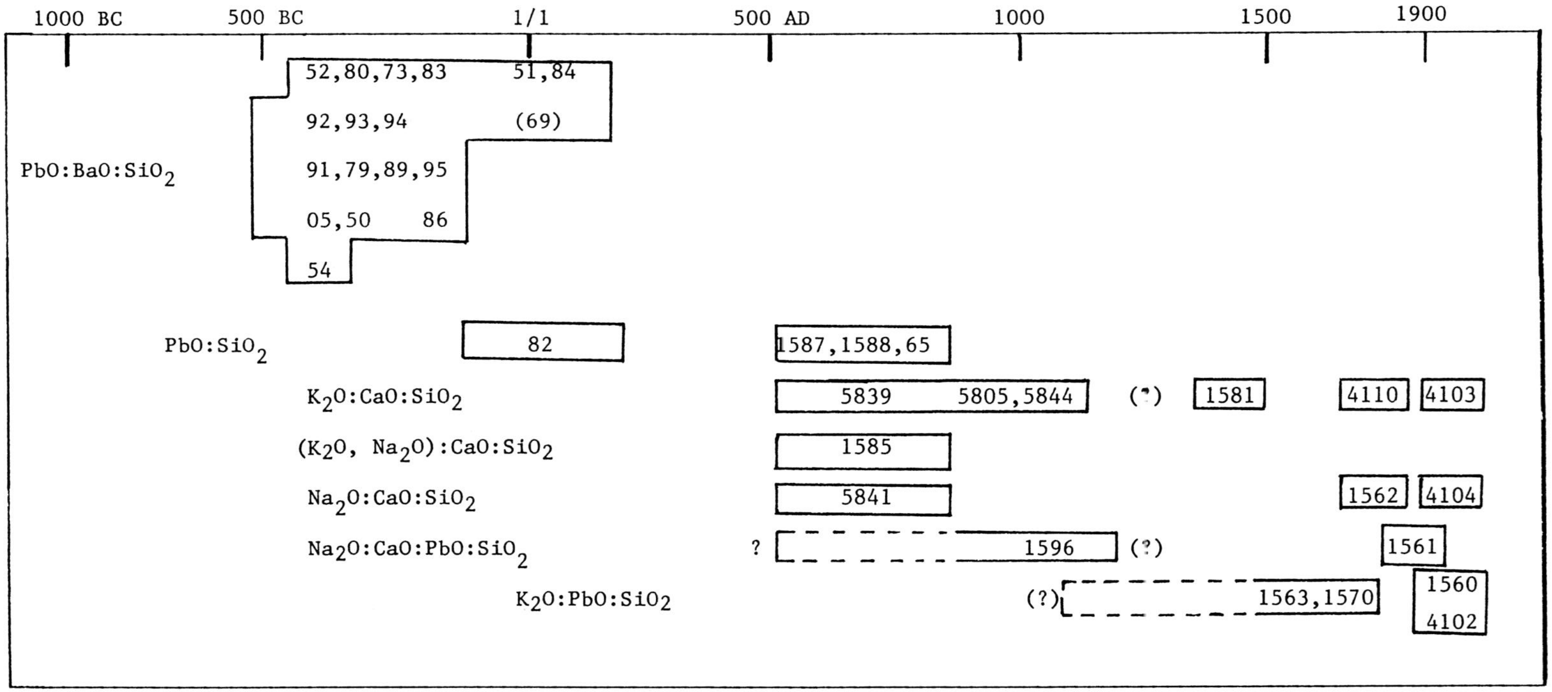

ADDENDUM TO CHAPTER 4

PHILIP M. FENN, ROBERT H. BRILL, AND SHI MEIGUANG

Corning, Inc., Corning, N.Y.; The Corning
Museum of Glass, Corning, N.Y.; and
China Building Materials Academy, Beijing

We present here quantitative chemical analyses of several additional samples analyzed while this book was in proof. They are samples nos. 3344, 3345, 3382, 3383, 5886, and 5892-94, descriptions of which are given below. The analyses are reported in Table 1.

Sample no. 3382 comes from a small cup, one of the first Han Dynasty glass vessels ever uncovered in an excavation. It has a squat, thick-walled, cylindrical shape and is made of light green, turbid glass, now heavily weathered. The cup was found in a tomb in Xuzhou, dated 175-128 B.C. In the same tomb, was found a large white opaque glass animal (no. 3383). Both glasses are heavily-leaded and contain barium, but they differ from one another in their relative proportions of PbO and BaO.[1]

Sample no. 3345 comes from an extremely interesting object, a glass vessel, now reconstructed from numerous fragments into the form of a beaker.[2] The rather thin walls, made of light blue opaque glass, are decorated with eye-like designs fused into place (see Fig. 1). Because of the resemblance of the decoration to that on eyebeads, the vessel is thought to date from the Warring States Period or the Han Dynasty. Those familiar with the Chinese glass in the Royal Ontario Museum will see a close similarity of the beaker to a group of six small fragments in that collection (see Figure 5 in Chapter 4). This similarity holds up even under microscopic examination. Both also contain a residue on the concave surfaces, probably left from their formation around a core of some sort. The appearance is reminiscent of the core residues on Egyptian cored vessels.[3]

Although the author who examined the beaker (R.H.B.) was unable to find any actual joins between the component parts of the beaker and the fragments in Toronto, we have nevertheless come to the conclusion that they might very well have come from the same parent object; if not, they undoubtedly came at least from the same workshop. This opinion is confirmed by two pieces of laboratory evidence. The first is the close similarity of the chemical compositions. An electron microprobe analysis of the beaker is reported in Table 1 here and an analysis of one of the Toronto fragments, no. 86, is reported in Table 1 of Chapter 4. They are a very close match, even to their somewhat higher than usual alumina contents. A microprobe analysis of the beaker glass by Julian Henderson[4] of Oxford was in close agreement with our own analyses. The second piece of evidence is the very close match of the lead isotope ratios determined for the light blue glasses of the cup and of the Toronto fragment. These are reported in Chapter 5 (no. 86) and the Addendum to Chapter 5 (no. Pb-2021).

Being inclined to believe that the pieces in question actually came from a single parent object, we would like to delve more thoroughly into their histories, so as to learn when and how they might have parted company. On the strength of the assumption that the glasses are from a single object, we have listed here the date of the beaker as the 4th-1st c. B.C., the same as that assigned by Doris Dohrenwend to the fragments in Toronto.

Table 1 also shows analyses of two rectangular plaques from a "glass burial suit", probably dating from the Han Dynasty (nos. 5892 and 5893). These pieces were lent to the Museum by Dr. Paul Singer for chemical analysis. They come from two groups of plaques which would, in part, have made up glass versions of the famous "jade burial suit". The glass plaques are illustrated in Figure 2.

Another "glass suit" from Yangzhou is described by Cheng Zhuhai and Zhou Zhangyuan in Chapter 2 of this book. That suit came from a late Western Han Dynasty tomb. The rectangular plaques in it appear to be a little longer and of slightly different proportions than the Singer plaques. The raised decoration on the Singer plaques might also be seen as representing the same sort of clouds, creatures, or plants as are depicted on the Yangzhou plaques. A microscopic examination of no. 5892 revealed traces of yellowish, blue, and green pigments on the decorated surface. No. 5893 shows traces of a red pigment resembling vermilion (cinnabar).

The two Singer plaques are lead-barium glasses. Their chemical compositions are so similar to one another that they could well have been made in the same place at about the same time. This view is strengthened by the fact that their lead isotope ratios are also very similar, indicating that they were very likely made with lead from a single, as yet unknown, source. (See Addendum to Chapter 5.) The plaques also resemble, in the same two ways, a *bi* and a dragon-decorated plaque from the Royal Ontario Museum (nos. 80 and 83 in Chapters 4 and 5). It is entirely possible that all four of these objects were made in the same factory, or at least in the same region.

But at the same time, the Singer plaques illustrate, too, one of the pitfalls encountered in interpreting analytical data on Chinese glasses. When comparing their analyses with that of the cup from Xuzhou (no. 3382), one sees a certain similarity in composition, and might, understandably, be led to conclude that the cup and plaques were made in the same place. However, lead isotope analyses (see no. 2012 in the Addendum to Chapter 5) show that the lead in the cup is markedly different from that in the plaques, and very likely came from a different mining region. Similarity in composition alone does not, apparently, necessarily imply that two glasses were made in the same place.

In connection with this, it is also worth noting that the analysis of the Yangzhou rectangular plaque, discussed above, is also a good match for the Singer plaques. Allowing for the fact that the analysis was carried out at a different laboratory, one might reasonably conclude that they could, for practical purposes, be nearly identical — but we do not have lead isotope data for the Yangzhou plaque, so there is still an element of uncertainty as to whether or not the Singer and Yangzhou plaques were made in the same place.

Sample no. 5886 is a truly intriguing, although not at all imposing, piece of glass. It is a small, slightly yellowish vessel with a raised interior ridge encircling a shallow kick. Coming from the Famensi in Shaanxi Province, the object was a companion, so to speak, of the scratch-decorated blue dishes found in a crypt beneath the pagoda. A repository of many extraordinary treasures, the crypt was sealed in 874 A.D. and is believed to have remained unopened until it was excavated a few years ago.[5] The blue dishes have not been analyzed, but two analyses have been done of this yellowish vessel, which might or might not have been made in the same place as the scratch-decorated blue dishes. Some examples of dishes bearing that sort of scratch decoration are thought to have been made in Nishapur, although others were definitely made elsewhere, possibly in Fustat.[6]

Two chemical analyses of separate chips of the yellowish dish were carried out using EDX[7] and the microbeam probe by S.M. and P.M.F. respectively. The results were in excellent agreement and are reported as a mean composition in the table under sample no. 5886. Except for the magnesia content (MgO = 5.41%), which is somewhat on the high side, the glass is a fairly good match for the mean composition of a group of twenty-three colored glasses from Nishapur.[7] On this basis, with the added information that the yellowish dish was found in probable association with the scratch-decorated blue dishes in the Famensi, we would tentatively attribute the yellowish dish to Nishapur — but that is an opinion which could be abandoned if compelling archaeological evidence suggested otherwise.

The final glass analysis in Table 1 is of a dark blue transparent, ellipsoidal bead purchased recently in Lhasa (no. 3344; see Figure 3). It is probably a piece of so-called "Peking glass". The analysis shows no surprises, corresponding to a rather straightforward soda-lime composition.

The remaining chemical analysis in Table 1 is that of a Chinese bead probably of the Han Dynasty (no. 5894). It is illustrated in Figure 4. This terra cotta bead, in the collection of Dr. Paul Singer, is coated with a bright blue, somewhat chalky, pigment. An x-ray diffraction pattern run by John Geiger of Corning, Inc. proved — as had been anticipated — that the pigment is Chinese blue (the tetrasilicate, $CuO.BaO.4SiO_2$). A semiquantitative emission spectrographic analysis, run by Nick Paliokastrites, also of Corning, Inc., confirmed the presence of Cu, Ba, and Si, and, in addition, 3-10% PbO. We expect that more and more examples of Chinese blue and of Chinese purple[8] will come to light as curators and archaeologists begin to look for them. A lead isotope analysis of the pigment on this bead is reported in the Addendum to Chapter 5.

Footnotes

1. These objects will be the subject of an article in a forthcoming issue of *The Journal of Glass Studies.*
2. *Inlaid Bronze and Related Material from Pre-Tang China, 11 June-5 July, 1991,* (an exhibition catalogue), Eskenazi, London, no. 69, p. 150; see also, pp. 21-24 for Julian Henderson's observations on the object. {Editor's Note: The analysis labeled "ROM glass", no. 4 in the table on p.23, is actually Corning's analysis reported in the Chinese version of the present book. The Eskenazi table contains some typographical errors. Also, we have revised the PbO and BaO values. For the correct and up-to-date data see no. 86 in Table 1 of Chapter 4 of this volume.}
3. R.H. Brill and J.F. Wosinski, "A Petrographic Study of Egyptian and Other Cored Vessels." *Proceedings of the VIIIth International Congress on Glass, London,* Committee B. Sheffield, England: The Society of Glass Technology, 1968, pp. 123-124.
4. Private communication.
5. Su Bai, "Famensi ta digong chu tu," *Wenwu,* 1988, pp. 29-43 (in Chinese); Robert H. Brill and Philip M. Fenn, "Some Thoughts on the Famen Temple Glass Finds," presented at the First International Symposium on Famensi Culture, Sept. 10, 1990, Fufeng County, Shaanxi Province. (In press.)
6. See Brill and Fenn cited in note 4.
7. R.H. Brill, "Chemical Analyses of Some Glass Fragments from Nishapur," in a forthcoming book on Nishapur finds by Prof. Jens Kröger of the Museum für Islamische Kunst, Berlin.
8. For information on these compounds see this volume, Chapters 4 and 5.

3382 Cylindrical cup; Xuzhou, Jiangsu Province, 175-128 B.C. Lt. green turbid glass, heavily weathered. Shi Meiguang no. WH-162. Sp. gr. 4.049. (Same as Pb-2012.)

3383 Large glass animal; same source as above. White opaque glass, lightly weathered. Shi Meiguang no. WH-166. Sp. gr. 3.85. (Same as Pb-2013.)

3345 Beaker-shaped vessel made up from fragments bearing eye motifs; prob. 4th-1st c. B.C. Lt. blue opaque glass with applied dots (lt. blue and black) over white and brownish fritted liners; moderately weathered. Inside retains core(?) residue. Lent by Eskenazi Ltd. to The Corning Museum of Glass for examination and analysis. This sample consists of lt. blue glass. (Same as Pb-2021.) Compare to CMG 86 thought to be from the same object.

5886 Famensi, Shaanxi Province; placed in crypt in or before 874 A.D. Fragment of a blown vessel, prob. a thin-walled, straight-sided cup, with an interior ridge surrounding the kick. P. yellowish glass, lightly weathered. Thought to resemble Famensi no. FD5 032.

3344 Medium-sized, flattened ellipsoidal bead; date uncertain, poss. "Peking glass". Dk. blue transparent glass, unweathered, but with wear. Purchased by R.H.B. in Lhasa, 9/30/90.

5892 Rectangular plaque from a "glass suit"; prob. W. Han. Perforations in corners. Raised, applied motifs suggest clouds, dragons, etc. Colorless turbid glass, heavily weathered. Darkened area of decoration contains a gummy substance and a finely divided white phase. Microscopic traces of yellowish, blue, and green pigments can be seen in a few small patches on the surface. Lent for sampling by Dr. Paul Singer. (Same as Pb-2028.)

5893 Rectangular plaque from a "glass suit"; prob. W. Han. Perforations in corners. No raised decoration, but preserves traces of red pigment. Colorless turbid glass, heavily weathered. Lent for sampling by Dr. Paul Singer. (Same as Pb-2029.)

5894 Cylindrical bead, terra-cotta with powdery, bright blue coating on surface; prob. Han. Length 2.6 cm, diam. 9 mm, bore 2 mm. Sample consists of blue material. XRD shows presence of Chinese blue, $\mathrm{CuO.BaO.4SiO^{2}}$. Lent for sampling by Dr. Paul Singer. (Same as Pb-2004.)

Figure 1. *Sample 3345. See also Figure 5 in Chapter 4. (Photo courtesy Eskenazi Ltd.)*

Figure 2. *Samples 5892 (top) and 5893 (bottom).*

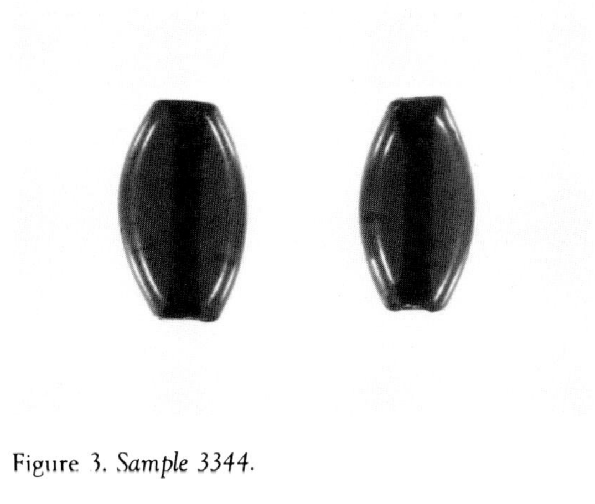

Figure 3. *Sample 3344.*

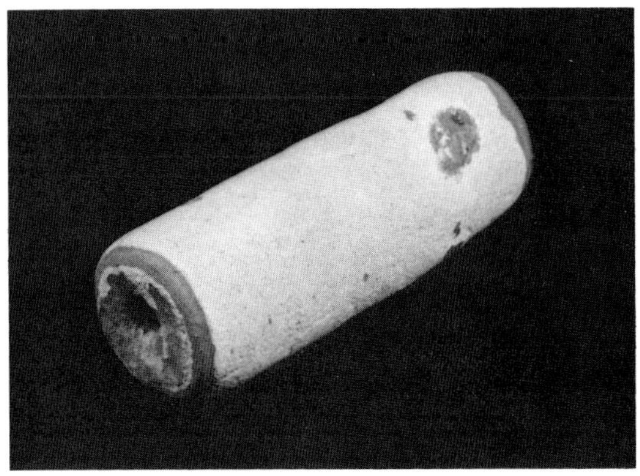

Figure 4. *Sample 5894.*

Table 1

Analyses of Some Additional Glasses

	175-128 BC *cup* *lt. grn. trbd.* 3382	175-128 BC *animal* *wt. op.* 3383	4th-1st BC *beaker* *lt. bl. op.* 3345	W. Han *plaque* *wt. trbd.* 5892	W. Han *plaque* *wt. trbd.* 5893	ca. 874 AD* *cup* *p. ylw.* 5886	19th-20th c.(?)** *bead* *dk. blue* 3344	prob. Han*** *Chin. Blue* *brt. blue* 5894
SiO_2	36.3	38.8	50.8	34.95	35.60	65.75	67.7	>30
Na_2O	2.19	3.34	1.49	2.82	2.70	15.08	17.9	0.3-1
CaO	0.16	0.71	2.56	0.10	0.10	5.12	7.29	0.3-1
K_2O	0.13	0.20	4.75	0.06	0.14	3.46	0.57	<1
MgO	0	0.14	0.58	0	0	5.41	0.44	0.1-0.3
Al_2O_3	0.37	1.28	6.29	.21	.24	1.95	4.26	1-3
Fe_2O_3	0.19	0.17	0.74	0.05	0.06	0.40	0.22	0.3-1
TiO_2	0.13	0.08	0.24	0.43	0.47	0.03	0.15	0.01-0.03
CuO	0.05	0.05	1.31	0.04	0.02	0.01	0.12	3-10
PbO	39.9	42.4	12.1	43.95	43.16	0.03	0.01	3-10
BaO	19.9	10.7	17.8	17.22	18.23	0.07	0.06	3-10
MnO	—	—	—	0	0	1.52	0.01	0.01-0.03
P_2O_5	—	—	—	0	0	0.13	0.05	<1

— Indicates not sought.

* Emission spectrographic analysis shows also 0.00X B_2O_3, and 0.00X V_2O_5. Sought, but not found, were: Sb, Co, Pb, Sn, Ag, Ni, and Zn.

** No. 3344 is colored with cobalt oxide.

*** Semiquantitative emission spectrographic analysis (on 0.5 mg sample) by Nick Paliokastrites and x-ray diffraction pattern identifying the pigment as $CuO.BaO.4SiO_2$ by John Geiger, both of Corning, Inc.

CHAPTER 5

LEAD ISOTOPE STUDIES OF EARLY CHINESE GLASSES

ROBERT H. BRILL, I. LYNUS BARNES, AND EMILE C. JOEL

The Corning Museum of Glass, Corning, N.Y.;
The National Bureau of Standards, Gaithersburg, Md.;
The Smithsonian Institution, Washington, D.C.

ISOTOPE analyses of lead extracted from ancient objects can be used to determine from which mining regions the leads could or could not have come. Consequently, the analyses are useful for learning more about where the objects themselves might have been made[1]. The method can be applied to ancient lead or to any other materials containing lead such as bronzes, silver, glasses, glazes, and pigments. Because isotope ratios are unaffected by chemical reactions or corrosion, weathering products are just as suitable for study as samples of unweathered materials. The analyses can be performed on samples which are so small as to be barely visible to the unaided eye. The analyses described here were carried out at the National Bureau of Standards, in Gaithersburg, Md., using our routine methods.[2] Previously, several hundred analyses had been made of ancient objects (mostly from sources other than Asia) and of galena ores.

In order to make use of the data, the isotope ratios determined for objects are compared with ratios for ores from mining regions known to have been worked in the past. There are two complications to be kept in mind. The first is the overlapping effect: lead ores from different regions sometimes have similar isotope ratios. Usually, however, some of the possible sources can be ruled out on the basis of archaeological considerations. The other complication is the mixing effect. If leads from different sources were salvaged and remelted together, then the isotope ratios of objects made from the mixture would be somewhere intermediate between those of the original leads.

Figure 1 summarizes the results obtained for several hundred samples in previous studies. For reference, the group marked "L" shows leads from the ancient mines at Laurion in Greece. Group E contains leads from England, Italy, Turkey, and perhaps elsewhere. Group S contains leads from Spain, Wales, and Sardinia. Leads found in early Egyptian glasses and glazes fall toward the lower left of the graph. A great many leads from Mesopotamia, and later samples from Egypt, fall in an unmarked region near Group L but above the general trend of the data as a whole.

Lead In Chinese Glasses

In an earlier study the lead in some Chinese glasses fell into two isotopic groups.[3] One group was toward the upper right corner of Figure 1, the other in the lower left. These groups contained the highest and lowest lead isotope ratios we have ever encountered in ancient objects from any part of the world. There are two significant aspects of this finding. First, the lead in the Chinese glasses is entirely different from that found in glasses made anywhere else in the world. Because ores with both of these unusual types of ratios are known to occur in China, this finding confirms that those particular glasses were made in China. Moreover, the two groups of Chinese leads are about as different as leads can be. Therefore, although some of these objects resemble one another closely in appearance and chemical composition, their leads must have come from mines which, geologically, were markedly different from one another. Similar observations were made on leads extracted from Chinese bronzes, although with the bronzes a greater number of samples, proportionately, fell in between the two extremes.

The present study adds fifty some additional specimens of glass and related materials to our body of data. The majority of the objects sampled (mostly fragments) are from the collections of The Royal Ontario Museum and The Corning Museum of Glass. They range in date from the 5th century B. C. to the early 20th century. Two glasses from private collections are included, as well as three beads excavated in Sumatra (by E. Edwards McKinnon) but thought to have been made in China.

The samples weighed only a few milligrams, being, for the most part, the remains left over

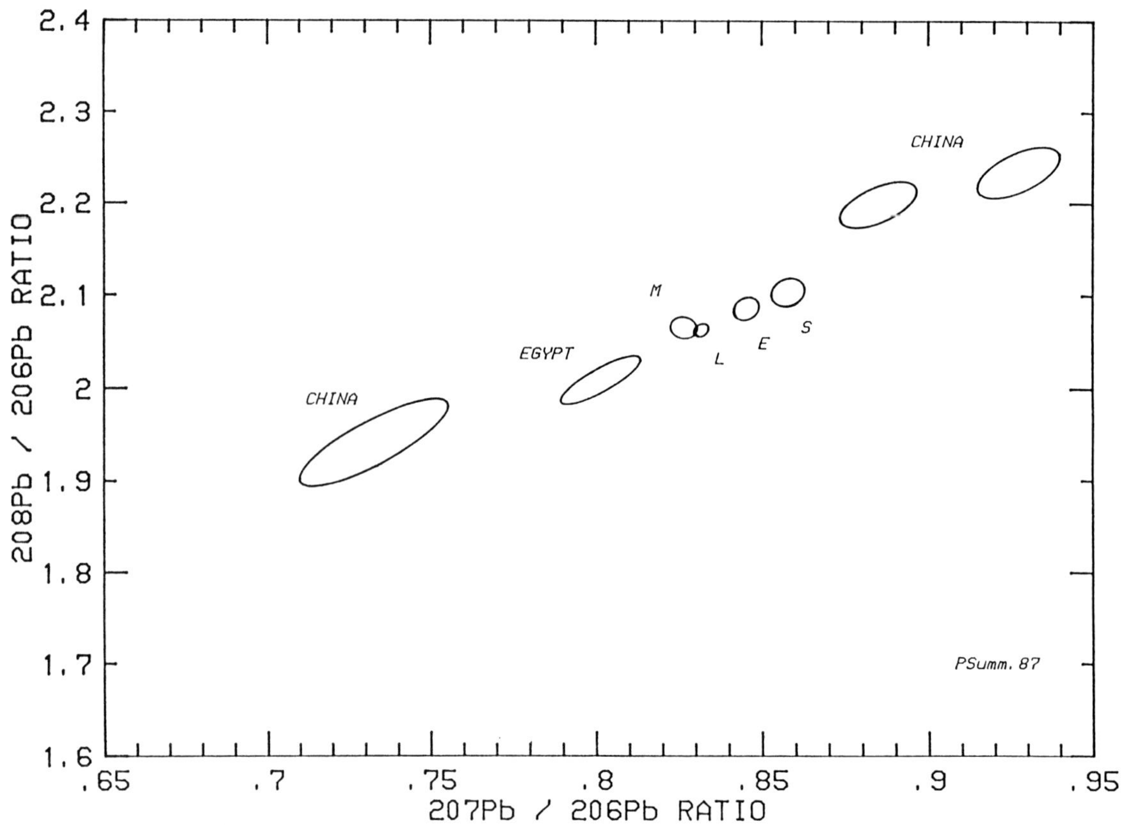

Figure 1. *Summary of lead-isotope ratios in archaeological objects and works of art from various places and times. All data run by the authors at the National Bureau of Standards.*

from sampling of objects for chemical analysis. In some cases they were the same samples used for refractive index measurements. The objects are described in the Catalogue of Samples where they have been divided chronologically into three broad periods labeled Early, Middle, and Late. Although each of these broad periods includes several of the dynasties and other periods into which Chinese cultural history is conventionally divided, the three broad designations do appear to separate the glasses according to their basic chemical formulations, and therefore they may also reflect differences in sources of lead used for glassmaking. Most of the objects are illustrated in another paper in this Symposium where chemical analyses of the same glasses are discussed.[4]

Results

Data for 62 glasses and related materials (and a few ores) are reported in Table 1 and plotted in Figures 2-4.[5] Figure 2 shows the results for the glasses of the Early Period (Warring States and Han); Figure 3 those of the Middle and Late Periods (Tang through Qing); and Figure 4 the ores and those objects which are either not glass or not Chinese. In Table 2 the data are listed in their approximate order of increasing ratios, and they are grouped so as to show which samples resemble one another most closely. Proxim-

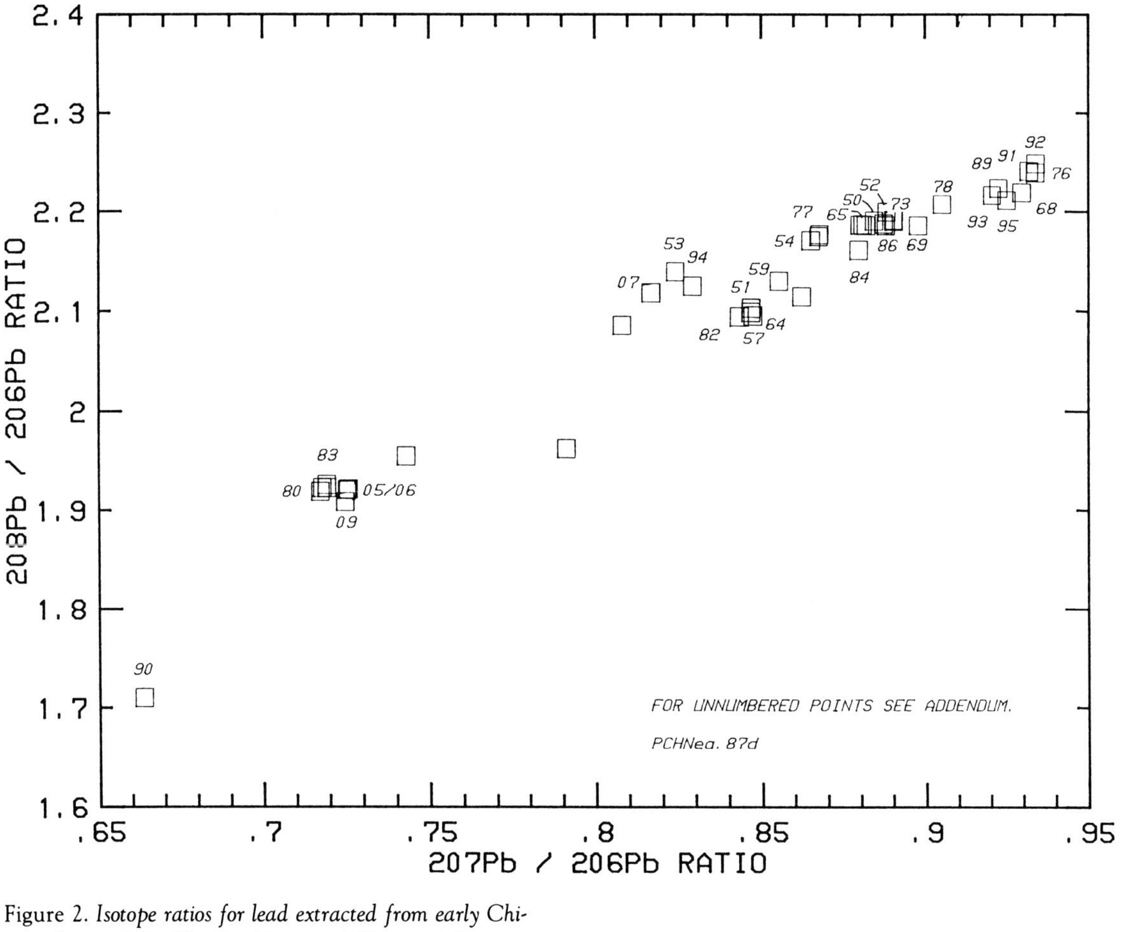

Figure 2. *Isotope ratios for lead extracted from early Chinese glasses of the Warring States and Han periods.*

ity in the list indicates similarity in ratios, and, therefore, likelihood of having had a common geographical origin. A single * separates similar, but not identical leads; ** represents a marked difference; *** denotes a very large difference. Table 3 shows how the glasses of different periods are distributed among ranges of isotope ratios.

The highest ratios determined were for the lead in a faceted "Peking glass" bracelet (no. 1561.) Its $^{207}Pb/^{206}Pb$ ratio is greater than 0.950. The lowest ratio was for the lead in the glaze of a faience eye bead of the 4-1st century B.C. Its $^{207}Pb/^{206}Pb$ ratio is just over 0.660. These extremes are the highest and lowest values we have ever found for any archaeological objects.

Early Periods

Among thirty-three glasses of the Early Period one finds again some samples with very low ratios, many with very high ratios, and now a few in mid-range. The Early glasses separate into five clusters with four or five samples in intermediate positions, and one lying well off by itself.

In some instances the isotope data group together samples having similar chemical compositions, but — somewhat surprisingly — there does not seem to be a strong overall correlation between the ratios and chemical composition. Perhaps we have just not yet found the best way of interpreting the chemical analyses. Some noteable examples of glasses where there is a strong correlation are: nos. 80 and 83; nos. 52, 65, 73, and 84; nos. 50 and 86; nos. 89 and 95; nos. 92 and 93.

Samples 05, 09, 80, and 83 are especially interesting because they form a rather tight group of low-ratio leads which probably have a common geographical origin. In addition, they are entirely different from the main body of the data. All four of these glasses are made of somewhat turbid, colorless glass and bear a certain stylistic resemblance.[6] The suggested dates, ranging from the 4-1st century B.C., are not firm, so the glasses could be more nearly contemporaneous than those dates suggest, and they are probably products of a single localized or regional industry.

Sample no. 90 is truly extraordinary, having the lowest ratios we have ever found in any archaeological object. The sample is from the very friable faience body of a 4-1st century B. C. bead with dotted decoration. The object with the closest stylistic and technological resemblance to this bead is no. 94, but that contains an entirely different type of lead.

Among the highest-ratio leads, there are, depending upon how one chooses to look at them, either one or two clusters of beads. We are inclined to see them as two closely-related but not identical clusters of three and four samples. All but two are colored, mostly blue, and only with copper, not with cobalt.

Sample 76, lying in the cluster of three, is significant because it is a yellow opaque glass. The sample was so small that it was not possible to perform even an emission spectrographic analysis to learn whether it was colored with antimony ($Pb_2Sb_2O_7$) or tin ($PbSnO_3$). The lead is clearly the same as that in the blue body glasses of two other beads and the glass must, therefore, have been made in China. Although not unexpected, it has until now remained unproven that the yellow glass was not imported from "the West." For comparison, in sample no. 56 the lead in the yellow opaque decoration on a large Near Eastern eye bead is quite different from the Chinese glass. It is plotted in Figure 4. We have found no Chinese objects containing this type of lead. The yellow opaque glasses deserve further careful study because the occurrence in China of this rather unusual synthetic material — one made by combining lead and antimony (or tin?) pyrotechnologically — implies the transfer of a highly specialized knowledge over great distances between China and the Near East. The alternative of two independent discoveries seems, to the authors, unlikely.

Moving down the graph past the eye bead no. 78 — or up from the bottom of Table 2 — another cluster of samples appears which, although still having high ratios, is distinctly different from those just described. This cluster contains nos. 52, 65, 69, 73, and 86. These are also eye beads, or objects closely resembling them in technique of manufacture, but some are colored with cobalt, not copper. This is an important glassmaking distinction suggesting a difference of time and/or place of manufacture. Allowing for the difference in color, the chemical compositions of the base glass of three of these samples, nos. 52, 65, and 73, and that of a near neighbor, a greenish turbid cicada, no. 84, turn out to be much like two of the glasses having very low lead ratios, nos. 80 and 83. This emphasizes that any connections between lead-isotope data and chemical analyses are not as straightforward as one might have hoped.

Entering the mid-range part of Figure 2 — or of Table 2 — there are two intermediate samples (nos. 54 and 77) and then two rather well-defined clusters. One contains a bead, a glass fish, a cicada, and an ear spool (nos. 51, 57, 64 and 82, respectively). Another cicada, no. 59, is nearby. Sample no. 82 is unusual because, although it has a high lead content, it does not have the same high barium content as do all the other glasses in this group for which we have chemical analyses. The type of lead characterizing these glasses is quite like that in a large group of glasses and other materials from the Shosoin in Nara, Japan. As will be mentioned below, this type of lead is also found in Middle and Late Period glasses and appears to be rather common in China. Although some overlapping may be at play here, we are inclined to believe that many, perhaps most, of these glasses must have a common origin, particularly the earlier ones.

Finally, the mid-range cluster of nos. 07, 53 and 94 seems to be displaced slightly above the general trend of the data, which might someday be an aid in matching it to lead ores. Sample no. 53 is the yellow opaque eye decoration in bead 52 whose dark-blue, cobalt-colored glass has a high-ratio lead. Thus the yellow glass was probably made at a different location than the blue base glass. It seems probable that in the case of the yellow glass the isotope ratios are telling us more about the source of the antimony (or tin) used to make the yellow colorant-opacifier than about where the glass itself was made. In this connection, it is significant that the lead in the yellow glass is a fairly good match for two other samples, an eye bead and a *bi* disk, the latter being of unquestionable Chinese origin. This places the antimony (or tin) source within China, and it all but rules out the possibility of the yellow pigment (or yellow cullet) having been imported as an item of commerce from the Near East.

Middle and Late Periods

The results for the five glasses of the Middle Period are plotted in Figure 3. The most striking observation about them is that none contains lead having either the low or very high ratios which characterize the earlier glasses. If this holds up as a generalization after further samples are analyzed, it will be of some importance. Such a finding implies that the lead in these glasses came from different mines than that in the earlier glasses and, therefore, that the glasses themselves probably came from different manufacturing centers. It could also simplify the interpretation of the findings on those earlier glasses which have the low and very high ratio leads.

Two Middle Period samples do, however, have ratios matching those of some of the high-ratio leads in one cluster of Early glasses. One is the small Buddha, no. 1581, of Tang date. Although only one example, it may indicate that glassmaking continued from the Han into the

Tang Period at the manufacturing center associated with that particular type of lead (or that it was revived there in the Tang Period.) Such a possiblility should be born in mind as more analyses become available. Another small Buddha, no. 1588, resembles the Buddha just mentioned, both stylistically and chemically, but its lead is quite different. Presumably, the two Buddhas were made in different places. No. 1587 is a much later object, a hairpin dated to the 15-16th centuries.

No. 1596, a large colorless bowl, matches the lead in a different cluster of Early glasses, one having mid-range ratios. It also overlaps three nearly identical leads, nos. 1560, 1563, and 1570, which are Late, and it is not very different from a late cameo glass (nos. 1566 and 1567). In view of the fact that the bowl's chemical composition reflects the use of rather pure raw materials, and because it also contains a small amount of arsenic, the bowl might be presumed to be of date later than Tang or Song. One cannot be certain (because of the overlapping effect), but lead of this mid-range ratio could conceivably have come from a single mining region which supplied glassmaking factories for as long as twenty centuries.

The five Late glasses, three of which are nearly indistinguishable in their isotope ratios, include a 20th century bead and two molded and

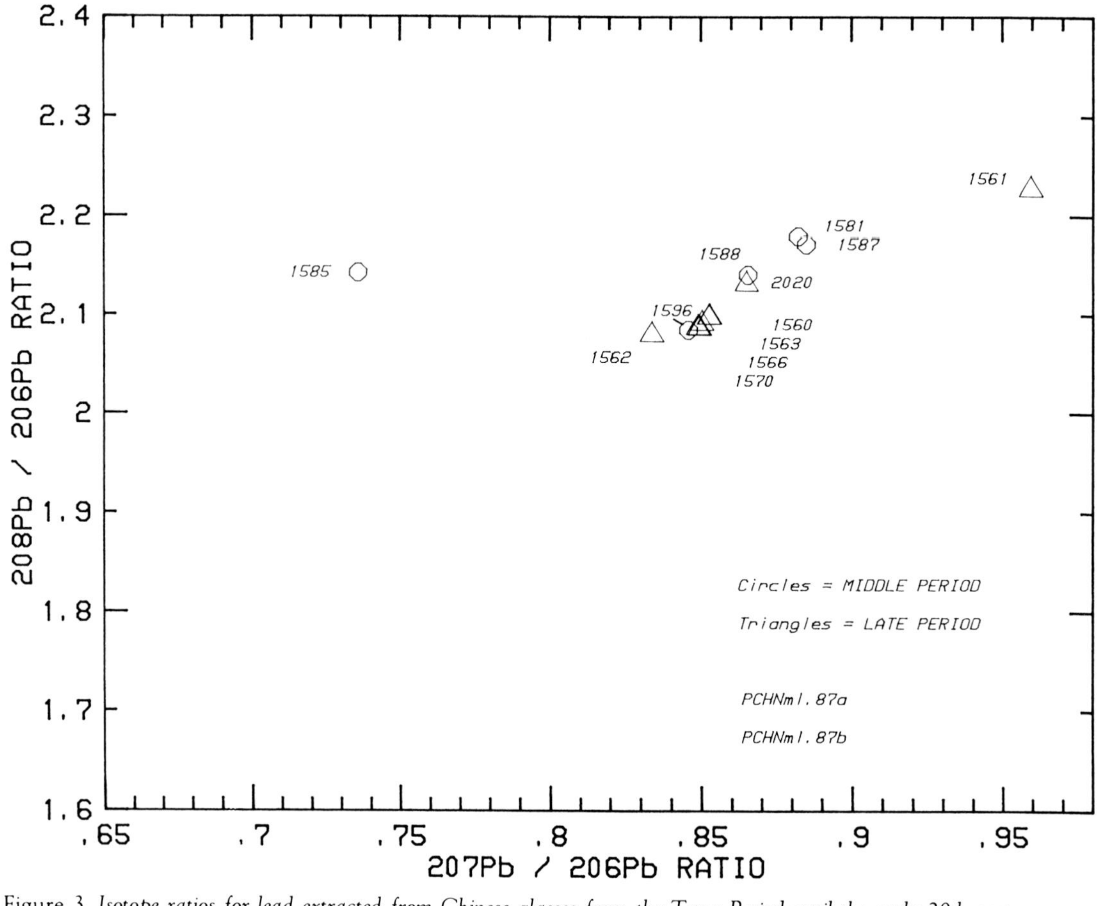

Figure 3. *Isotope ratios for lead extracted from Chinese glasses from the Tang Period until the early 20th century.*

cut figurines. The two figurines (nos. 1563 and 1570) are so close in both isotope ratios and chemical composition that they must bear a closer relationship than their assigned dates suggest. On the basis of that reasoning, one is inclined to question the early limit (12th century) of the dating of the Guan Yin (no. 1570). Perhaps it is better to date both of them to a range of the 17-19th centuries. They are done in distinctly different styles, but one wonders if the manufacturing center where they were made might not have been making figurines for different markets. The Buddha (no. 1563), although actually found in Thailand, is believed to have been made in China. (See the reference in Catalogue of Samples.) The blue and white glasses (nos. 1566 and 1567) from a "cameo-carved" vessel, probably of the 19th century, contain leads closely resembling those in the Buddhas. Another Late glass, no. 1562, a badly crizzled blue bottle of the 18-19th centuries, differs somewhat from those just discussed. No. 1585, a Tang hairpin head, has very unusual isotope ratios unlike any we have encountered before. Perhaps it was made at some factory well outside the mainstream of glassmaking in China.

Other Materials

Eight supporting samples, which are either not glass or not Chinese, are plotted in Figure 4. The most interesting are nos. 97, 2000, and 2001. These are bars of "Chinese blue" or "Chinese purple," dated to the Han Dynasty. The material has been found in the form of small sintered and molded bars as well as a pigment painted on Chinese vessels. It is newly-identified, a Chinese analogue of Egyptian blue. Egyptian blue is a man-made material known from throughout Egypt, the Near East, and Europe from ca. 1500 B.C. until Roman times, and possibly as late as Mediaeval times. Egyptian blue is a definite chemical compound, $CuO.CaO.4SiO_2$. Zycherman and FitzHugh recently described Chinese blue and identified it as $CuO.BaO.4SiO_2$.[7] We have confirmed their findings by electron microprobe analyses and x-ray diffraction, and we have prepared another new compound having an empirical formula close to $CuO.BaO.2SiO_2$. This has a purple color and hence has been named "Chinese purple." The blue tetrasilicate and the purple disilicate (if that is what it is) are different compounds, each having its own x-ray diffraction pattern. We have synthesized them by the same two methods we used for synthesizing Egyptian blue: by heating dry batch materials and by devitrification of glasses. The purple phase forms from some batches at about 1035-1080°C, and the blue phase at somewhat higher temperatures ($\sim$1080-1160°). The temperatures at which the two phases form appears to depend upon the exact composition of the batch. Both phases, it seems, occur in the Han Dynasty bars analyzed here. The most significant point is that in the Chinese version of this pigment BaO has replaced CaO, just as it did (in a loose sense) in Chinese glasses of about the same period. The most significant question raised is whether Chinese blue and Chinese purple were independently discovered in China or whether they stemmed from some transfer of technology from civilizations to the west.

The three samples of Chinese purple contain minor percentages of lead. The lead in one of the samples (no. 97) resembles those in some BaO-containing glasses with mid-range isotopic values. The other two (nos. 2000 and 2001) are nearly identical to one another and to the leads in one of the high-range groups of eye beads and other glasses containing BaO. Most likely, all were made in the same region.

No. 08 is from a yellow ceramic, known to have been made in Jingdezhen, and it offers a starting point for identifying leads associated with that famous ceramic center. No. 56 is the Near Eastern bead discussed above with the Chinese yellow opaques. Its isotope ratios match those of a group of other leads (including some yellow glasses and glazes) from Persia and Mesopotamia. No. 72 is a bronze buckle of uncertain date containing two blue glass inlays which could not be sampled. Its isotope ratios match those of the mid-range cluster described just above, that containing glasses from several periods. Two glazes from architectural tiles at Tanzhe Monastery (nos. 2010 and 2011) match a group of mid-range early glasses. More significantly, however, these tiles, which could date anywhere from the Qing Dynasty to the 19th century, also are a good match for the lead in a 19th century cameo glass (nos. 1566 and 1567). Finally, no. 1599, a green glaze from a terra cotta animal with a Central Asian or Chinese look about it (but acquired in Ghazni, Afghanistan) also has that same kind of lead.

The picture is considerably brighter with respect to those Early glasses which have low, high, or very-high ratios. Among the data of Chen *et al.*,[8] relatively few ores were reported which contain these isotopic types of galenas. Also, the seven Chinese galenas we have analyzed (Figure 4) all have mid-range ratios. The fact that so many of the Early glasses contain leads lying at the extreme limits of the galena ranges greatly strengthens their specificity. The lowest $^{207}Pb/^{206}Pb$ ratio reported by Chen *et al.* is 0.7624, whereas the low ratio values found in the glasses lie below that limit, ranging between 0.710 and 0.730. That is a large isotopic gap. The highest ore values they report are about 0.883 and 0.885. These values just manage to include a few of the high-ratio glasses analyzed, and there are twelve Early glasses which have values greater than that limit. In fact, the seven very-high ratio glasses have values greater than 0.910, well above the highest Chinese ores yet reported. The point of all this is that the isotope ratios of the leads found in the Early glasses appear to be very rare among galena ores occurring naturally in China. Therefore, there must be relatively few galena deposits which could have provided the lead actually found in the glasses, meaning that once those deposits are located, there should be little doubt but that they really supplied the lead in the glasses. Archaeological and historical reasoning should narrow down the possibilities even further.

The highest values reported by Chen *et al.* are from Liaoning (Fengchen), Jilin (Jian, Hanwu), and Xinjiang (Kalanguguole), while the lowest are from Yunnan (Huize, Yongshan). Within the range 0.810-0.840 are ores from Anhui, Guizhou, Guangxi, Yunnan, and Xizang (Tibet). One might look in those places, or in places with similar geological formations, for ores matching glasses in the cluster containing glasses nos. 07, 53 and 94.

The mid-range cluster at about 0.840-0.850, containing glasses of all three periods, matches ores from several places in southern and eastern China and from Gansu and Ne Meng Gu (Inner Mongolia). This may be the overlapping effect at its worst regarding Chinese glasses.

Three Glasses from Sumatra

We also analyzed a white opaque glass bangle (no. 2076) and two yellow opaque glass beads (nos. 2077 and 2078) excavated at Kota Cina by E. Edwards McKinnon.[9] Nos. 2076 and 2078 resemble one another in their isotopic compositions, but no. 2077 is somewhat different. Although found in Sumatra, the objects were accompanied by Chinese ceramic sherds. This, along with the fact that all three beads are heavily-leaded,[10] led us to believe that they might have been imported from China. Their $^{207}Pb/^{206}Pb$ ratios fall among the mid-range Chinese leads and are similar, although not identical, to the two glass figurines mentioned above (nos. 1563 and 1570), the Tang bowl (no. 1596), the two glazes from the Tanzhe Monastery (nos. 2010 and 2011), the cameo-carved vessel (nos. 1566 and 1567), and a 20th-century bead (no. 1560). However, there is a significant spread among the $^{208}Pb/^{206}Pb$ ratios suggesting that geologically dissimilar ores provided the leads in these glasses. Galena ores having ratios resembling those of these glasses occur in several places throughout Guangdong and in Hunan,[11] including Shuigoushan which is still a major mining region today. Overlapping may again come into play here. An unpublished analysis of a galena from Malaysia by one of the authors (I.L.B.) is quite similar to one of the Guangdong galenas (no. 1804).

Prospects for Further Research

In the discussions above we have looked mainly at the $^{207}Pb/^{206}Pb$ ratios plotted along the horizontal axes. It should be noted, however, that the points do not tend to generate a single linear trend, but, instead, seem to be generating a family of more-or-less parallel trends. Another way of looking at this is to say that some samples show a spread in their $^{208}Pb/^{206}Pb$ ratios which places them above or below some general trend. Given the accuracy and precision of the analyses,[12] this effect is certainly real and is consistent with what has been observed in previous studies. In a positive sense, this can be seen as reducing overlapping, thus paving the way for more specific classifications of objects and improving the prospects for matching samples with one another and with particular mining regions. In any event, we are still some way from being able to accomplish that goal for Chinese glasses — particularly those with mid-range values.

Archaeometric publications often conclude with a remark to the effect that "further data are required." Experience tells us, however, that all too often further data only serve to muddy a picture rather than to clarify it. In this instance, for reasons which in truth might be subjective, the authors really do take an optimistic view. What will ultimately be required are reliable, high-accuracy ore data, a sorting out of which ore deposits could reasonably be expected to have been worked in ancient times, and analyses of as many glass samples as possible. This should then be followed by a careful and objective review of all the relevant data, both isotopic and archaeological.

Notes

1. R. H. Brill and J. M. Wampler, "Application of Science," *Examination of Works of Art*, Museum of Fine Arts, Boston, 1965, p. 155; R. H. Brill and W. R. Shields, *Special Publication No. 8*, Royal Numismatic Society, University Press, Oxford, 1972, p. 279; R. H. Brill, I. L. Barnes, and B. Adams, "Lead Isotopes in Some Ancient Egyptian Objects," *Recent Advances in Science and Technology of Materials*, 3, Plenum Press, New York and London, 1974, pp. 9-27; I. L. Barnes, W. R. Shields, T. J. Murphy, and R. H. Brill, "Isotopic Analysis of Laurion Lead Ores," *Archaeological Chemistry*, Chapter 1, American Chemical Society, Washington, D.C., 1974, pp. 1-10; I. L. Barnes, R. H. Brill, E. C. Deal and G. V. Piercy, "Lead Isotope Studies of Some of the Finds from the Serce Liman Shipwreck," in *International Archaeometry Symposium*, ed. Jacqueline S. Olin, M. James Blackman, Smithsonian Institution Press, 1986, pp. 1-12.
2. I. L. Barnes, T. J. Murphy, J. W. Gramlich, and W. R. Shields, *Analytical Chemistry*, v. 45, no. 11, 1973, pp. 1881-84.
3. R. H. Brill, K. Yamasaki, I. L. Barnes, K. J. R. Rosman, and M. Diaz, "Lead Isotopes in Some Japanese and Chinese Glasses," *Ars Orientalis*, v. 11, 1979, pp. 87-109.
4. R. H. Brill, S. C. Tong, and D. Dohrenwend, "Chemical Analyses of Some Early Chinese Glasses." This volume, Chapter 4.
5. The $^{204}Pb/^{206}Pb$ data were plotted but, as is usually the case, they did not produce any additional information useful for classifying the samples.
6. See illustrations in this volume, Chapter 4.
7. See this volume Chapter 4 and Addenda to Chapters 4 and 5.
8. Chen Yuwei, Mao Cunxiao, and Zhu Bingquan, *Geochimica*, v. 3, no. 9, 1980, p. 215. (In Chinese.) See also, this volume, Chapter 16.
9. E. Edwards McKinnon and R. H. Brill, "Chemical Analyses of Some Glass from Sumatra," *Archaeometry of Glass, Archaeometry Session of the XIVth International Congress on Glass, New Delhi, 1986*, Indian Ceramic Society, Calcutta, Section 2, 1987, pp. 1-14.
10. The full analyses are reported in the reference cited in note 9.
11. For example, our nos. 1804 and 1805, as well as several ores listed in Chapter 16.
12. See references cited in note 1.

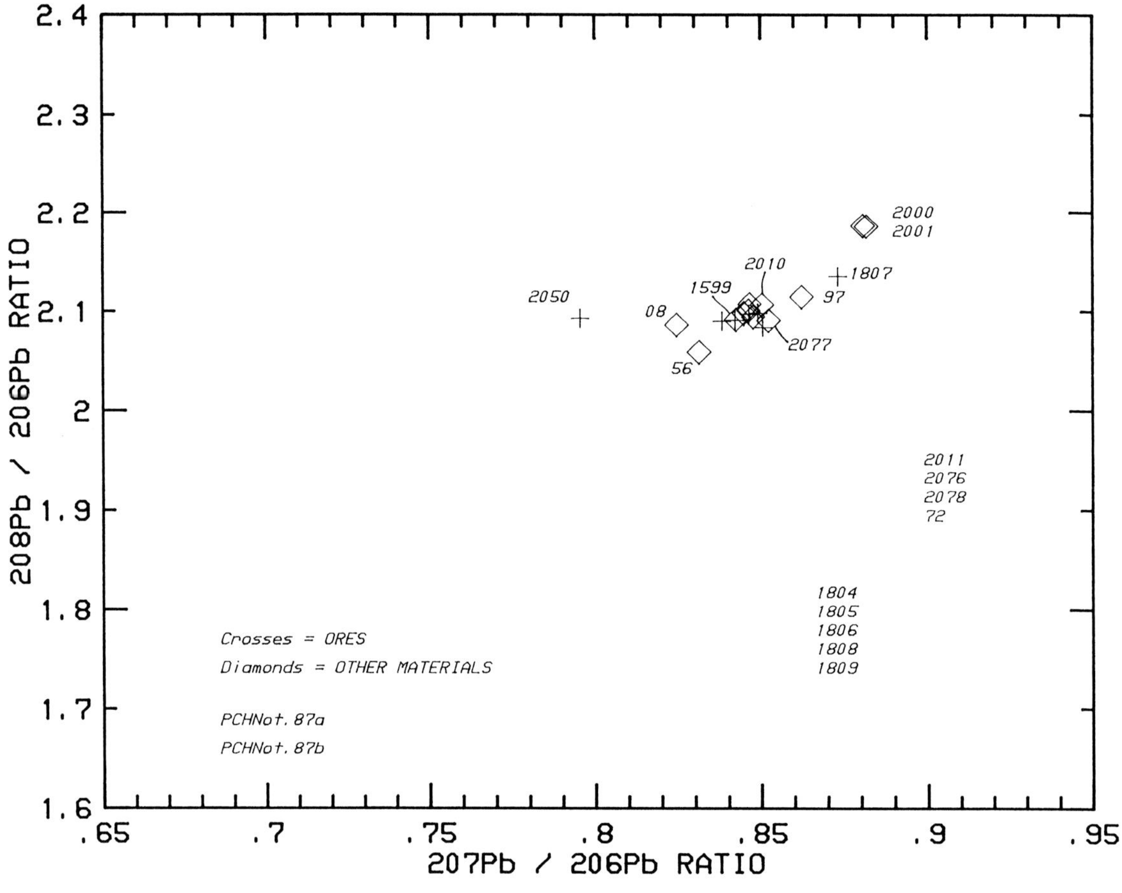

Figure 4. *Isotope ratios for lead extracted from some Chinese ores and miscellaneous materials.*

Table 1

Isotope Ratios of Lead in Chinese Glasses and Related Materials

Sample number	*NBS number*	$^{208}Pb/^{206}Pb$	$^{207}Pb/^{206}Pb$	$^{204}Pb/^{206}Pb$
Early Periods, Glasses				
05	694	1.9211	0.7257	0.04516
06	695	1.9203	0.7253	0.04509
07	696	2.1178	0.8168	0.05167
09	607	1.90812	0.72467	0.045114
50	700	2.1902	0.8842	0.05693
51	701	2.1026	0.8468	0.05372
52	1084	2.19931	0.88811	0.057455
53	1085	2.13889	0.82403	0.052270
54	1083	2.17023	0.86497	0.055556
57	1086	2.09487	0.84730	0.053941
59	1087	2.12944	0.85531	0.054843
64	1375	2.09833	0.84665	0.053828
65	1805	2.18596	0.87983	0.056662
68	1396	2.21921	0.92968	0.060933
69	1397	2.18530	0.89767	0.058368
73	1290	2.18996	0.89034	0.057612
76	1398	2.23950	0.93384	0.061283
77	1399	2.17426	0.86731	0.055817
78	1400	2.20710	0.90507	0.059169
80	1272	1.92226	0.71780	0.044405
82	1273	2.09396	0.84332	0.053581
83	1274	1.92554	0.71917	0.044547
84	1275	2.16063	0.87950	0.056668
86	1276	2.19092	0.89024	0.057482
89	1279	2.22356	0.92251	0.060179
90	1280	1.70979	0.66357	0.039786
91	1281	2.24107	0.93190	0.061050
92	1282	2.24866	0.93394	0.061274
93	1283	2.21637	0.92038	0.060377
94	1284	2.12441	0.82924	0.052679
95	1285	2.21130	0.92502	0.060697
Middle Periods, Glasses				
1581	1292	2.17914	0.88218	0.056825
1585	1291	2.14216	0.73558	0.046237
1587	1277	2.17102	0.88484	0.057149
1588	1278	2.14043	0.86534	0.055403
1596	1411	2.08431	0.84564	0.054077

Late Periods, Glasses				
1560	1089	2.09112	0.85017	0.054334
1561	1412	2.22774	0.95942	0.062213
1562	1096	2.07958	0.83366	0.053051
1563	1287	2.08633	0.84941	0.054317
1566	1554	2.09750	0.85245	0.054405
1567	1555	2.09777	0.85275	0.054407
1570	1288	2.08676	0.84883	0.054218
Other				
08	606	2.08593	0.82448	0.052394
56	1088	2.05848	0.83119	0.053042
72	1401	2.10024	0.84622	0.054063
97	1293	2.11394	0.86222	0.055077
1599	1286	2.09117	0.84226	0.053617
2000	1556	2.18556	0.88169	0.056758
2001	1557	2.18623	0.88061	0.056769
2010	1924	2.10599	0.85030	0.054058
2011	1925	2.09401	0.84758	0.053992
2076	1704	2.09762	0.84470	0.053659
2077	1705	2.09043	0.85227	0.054607
2078	1706	2.10669	0.84650	0.053770
Ores				
1804		2.0964	0.8463	0.05369
1805	711	2.0975	0.8490	0.05422
1806	1392	2.09051	0.84192	0.053418
1807	1393	2.13448	0.87312	0.056090
1808	1394	2.08972	0.83793	0.053517
1809	1395	2.08343	0.85050	0.054816
2050	1565	2.08232	0.84115	0.053977

Table 2

Samples Grouped by Isotope Ratios

EARLY PERIODS		MIDDLE, LATE PERIODS	
Low ratios (<0.800)			
90	Eye bead (faience), BaO (?)		
***			
80	*Bi*, BaO		
83	Plaque, BaO		
05, 06	*Bi*, BaO		
09	Sword terminal, BaO (?)		
***			
Mid-range ratios (0.800-0.870)			
07	*Bi* (na)		
53	Yellow inlay from 52 (na)		
94	Eye bead (terra cotta) BaO		
		**	
*		1562	Bottle, 18-19th c.
		*	
		1596	Bowl, Tang/Song
51	Bead, BaO	1560	Bead, 20th c.
64	Cicada (na)	2010, 11	Glazes, 17-19th c.
57	Fish (na)	1566, 67	Cameo, 19th c.
82	Ear spool, no BaO	1563	Buddha, 17-19th c.
59	Cicada (na)	1570	Guan Yin, 12-17th c.
97	Chinese purple		
		*	
*		1588	Buddha, 7-8th c.
		*	
54	Eye bead, BaO		
77	Bead, faience		

Table 2 (continued)

High ratios (0.870-0.910)

*			
84	Cicada, BaO	*	
		1581	Hairpin, 15-16th c.
*(?)		1587	Buddha, 7-8th c.
65	Eye bead, BaO		
2001	Chinese purple		
2000	Chinese purple		
50	Eye bead, BaO		
52	Eye bead, BaO		
73	Crescent, BaO		
86	Vessel (?), BaO		
69	Eye bead, BaO		
*(?)			
78	Eye bead (na)		

Very high ratios (>0.910)

**			
93	Bead, BaO		
89	Bead, BaO		
95	Eye bead, BaO		
68	Eye bead, BaO		
*(?)			
91	Eye bead, BaO		
92	Bead, BaO		
76	Eye bead, yellow (na)	1561	Bracelet, 19-20th c.

(na) not analyzed

Table 3

Distribution of Periods by $^{207}Pb/^{206}Pb$ Ratios

Low (<0.800)	
	Early periods only (6 glasses)‡
Mid-range (0.800-0.870)	**
	Early (3)
	*
	Late (1)
	*
	Early (5) Middle (1) Late (7)
	*
	Early (2) Middle (1) Chinese purple (1)
High (0.870-0.910)	*
	Early (8) Chinese purple (2) Middle (2)
Very high (>0.910)	**
	Early (7) Late (1)

‡ No. 1585, a Middle Period glass, has an unusual isotopic composition and contains only a trace of lead.

CATALOGUE OF GLASS SAMPLES

Note: This catalog, although it repeats some information from that in Chapter 4, is included for the reader's convenience. Many of the objects described here are illustrated in Chapter 4.

Studies completed:
- * quantitative analysis (aa)
- e quantitative analysis (microbeam)
- s qualitative analysis (spectrography)
- \+ Pb isotope analysis
- p some physical properties

Early Periods

*	s	+	05	*Bi* disk, 4th-3rd century B.C. Colorless glass, heavily weathered. CMG 51.6.567. (3330, Pb-1005)
*	s	+	06	Weathering product from 05. (3331, Pb-1006)
		+	07	*Bi* disk, 4th-3rd century B.C. (?). Turbid green glass, lightly weathered. CMG 51.6.548. (Pb-1007)
*	s	+	09	Sword scabbard terminal, 2nd century B.C. Slightly turbid colorless glass, heavily weathered. CMG 62.6.15. Emission spectrography shows 25% PbO; 0.4% BaO. (3229, Pb-1009)
*	s	+	50	Eye bead fragment, 4th-3rd cent. B.C., possibly from Jincun near Luoyang. Dark blue transparent glass with white opaque decoration marvered in, moderately weathered. CMG 51.6.549B. (3334, Pb-1550)
*	s	+	51	Large bead, Han (?). Light blue transparent glass apparently undecorated, heavily weathered. CMG 67.6.2. (3342, Pb-1551)
*	s	+	52	Eye bead, 4th-3rd cent. B.C. Dark blue transparent glass with yellow opaque and white opaque decoration marvered in, very lightly weathered. (3335, Pb-1552)
		+	53	Yellow opaque glass from same object as 52. (Pb-1553)
*	s	+	54	Eye bead, 4th-3rd cent. B.C. Dark blue transparent glass with inlaid white (?) opaque and yellow (?) opaque decoration, moderately weathered. (3336, Pb-1554)
		+ p	57	Glass fish, flat with incised decoration, probably Han. White opaque glass, moderately weathered. CMG 51.7.3. (3338, Pb-1557)
		+	59	Glass cicada, Han. Greenish glass, heavily weathered. Submitted by Rutherford J. Gettens, Freer Gallery of Art SC-543. (240, Pb-1559)
	s	+	64	Glass cicada, Han. Turbid white glass, heavily weathered. CMG 51.6.558. (3341, Pb-1564)
		+	65	Eye bead, Warring States, Honan Province. Dark blue transparent glass, lightly weathered. Submitted by Zhang Fukang, his no. G-41. 33.92% PbO; 20.05% BaO, colored with CoO; MnO not reported. (Pb-1565)
	s	+	68	Eye bead, Han (?). Light blue transparent glass, moderately weathered. Beck Collection "C95317, sp. gr. 3.27." (5867, Pb-1568)
*	s	+	69	Eye bead, Han (?). "Black," frit-like body with light blue, turbid inlay eye and remains of white fill. Beck Collection "CGS, E, B/2455, sp. gr. 3.1." (5864, Pb-1569)

e +	73	Crescent with eye decoration, 4th-1st cent. B.C. Dark glass, heavily weathered. ROM 933.12.167. (5808, Pb-1573)
+	76	Eye bead, 5th-1st cent. B.C. Blue (?) glass with decoration marvered in, heavily weathered. ROM 933x84.1. Sample is of yellow opaque glass. (5816, Pb-1576)
+	77	Glazed bead, 4th-1st cent. B.C. Frit-like body with weathered green glaze, "S" and circle decoration. ROM 933.84.11. (5826, Pb-1577)
+	78	Eye bead, 4th-1st cent. B.C. "Black" glass with inlay eyes and aqua glass remaining in some places and double bands of dotted lines once containing a white fill; moderately-heavily weathered. Sample is of black glass. ROM 933.84.7. (5818, Pb-1578)
* s e + p	80	*Bi* disk, broken and repaired, 3rd cent. B.C. Slightly turbid (?) pale green glass, heavily weathered. ROM 971.144.4. (5830, Pb-1580)
* s e + p	82	Ear spool, Han. "Black" glass, heavily weathered. ROM 933.84.13. (5832, Pb-1582)
* s e + p	83	Glass plaque with dragon, 2nd-1st cent. B.C. Turbid colorless glass, heavily weathered. ROM 933.23.340. (5835, Pb-1583)
* s e + p	84	Glass cicada, Han. Turbid pale greenish glass, heavily weathered. ROM 930x170.9. (5837, Pb-1584)
* s +	86	Vessel (?) fragment, 4th-1st cent. B.C. Light blue opaque glass with applied dots of same color set over white (?) liners, moderately weathered. Underside may contain traces of resinous material. ROM 933.58.5 (5842, Pb-1586)
* s +	89	Barrel-shaped bead, 4th-1st cent. B.C. Reddish-brown glass body, (original color indeterminate), with applied eyes of light blue translucent glass, heavily weathered. ROM 933.12.165. (5848, Pb-1589)
s +	90	Eye bead, 4th-1st cent. B.C. Friable gray body of pumice-like appearance; glazed surface with dotted decoration of indeterminate original color, heavily weathered. Sample is of body material. Emission spectrography showed glaze to contain some lead. ROM 960.241.248. (5851, Pb-1590)
* s e + p	91	Eye bead, 5th-1st cent. B.C. Medium blue glass body, inlay eye decoration missing, heavily weathered. ROM 933.84.3. (5852, Pb-1591)
* s + p	92	Bead, 5th-1st cent. B.C. Greenish-blue glass, heavily weathered. Might once have contained eye decoration or might have been faceted. ROM 933.84.4. (5854, Pb-1592)
* s + p	93	Small bead, 4th-1st cent. B.C. Black glass, probably once contained eye decorations; heavily weathered. ROM 933x84.5. (5855, Pb-1593)
* s +	94	Eye bead, 4th-1st cent. B.C. Friable terra cotta (?) body, black glazed surface, traces of green fill (probably glass) and white liner remain, also dotted decorations of white fill, heavily weathered. Sample is of black glaze. ROM 933x84.6. (5856, Pb-1594)
* s + p	95	Cube-shaped bead with raised eye decorations at each corner, 4th-1st cent. B.C. Colorless glass with blue glass eyes; heavily weathered. ROM 933x84.8. (5858, Pb-1595)

Middle Periods

* s + p 1581	Glass hairpin, 15th-16th cent. A.D. Light blue opaque glass, lightly iridescent in places. ROM 960.241.230. (5831, Pb-1581)
* s e + p 1585	Hairpin head, molded relief design, 7th-9th cent. A.D. Somewhat turbid colorless glass, moderately weathered. ROM 930.170.3. (5840, Pb-1585)
* s + p 1587	Small glass Buddha, 7th-8th cent. A.D. Amber glass, moderately weathered. ROM 925.26.367A. (5845, Pb-1587)
* s + p 1588	Small glass Buddha, 7th-8th cent. A.D. Light green opaque glass, moderately weathered. ROM 925.26.367L. (5846, Pb-1588)
* s + p 1596	Large bowl, Tang-Ming or 8th-17th cent. A.D. Said to be from tomb in temple ground at Luoyang. Colorless transparent glass with brownish substance and rough grayish white deposit on parts of interior. Diameter 21.3 cm. ROM 923.24.163. (5860, Pb-1596)

Late Periods

* s + p 1560	Large glass bead, probably early 20th cent., acquired in China. Possibly made in China but more likely imported from Europe. Bright green transparent glass colored with uranium, unweathered. Diam. 2.8 cm, drilled bore. CMG 51.6.246. (4100, Pb-1560)
* s + 1561	"Peking" glass bracelet, faceted, 19th or early 20th cent. Dark blue transparent glass, incipient crizzling (?) CMG unaccessioned. (4108, Pb-1561)
* s + 1562	Bottle, 18th-19th cent. Dark blue transparent glass, crizzled with surface spalling away. Submitted by Axel von Saldern, Museum für Kunst und Gewerbe, Hamburg. (4111, Pb-1562)
* s e + p 1563	Glass Buddha, mold cast with engraved details, 17th-19th cent., probably Thailand. Gray transparent glass with loose, brownish accretion. See A.B. Griswold, "A Glass Image and Its Associations," *Journal of Glass Studies*, v. 5, 1963, pp. 75-103. CMG 56.6.10. (5870, Pb-1563)
* s + 1566	Cameo-carved vessel, 19th cent. White opaque base glass. Submitted by Robert H. Clague. (4113, Pb-1566)
* s + 1567	Same object as 1566. Dark blue transparent overlay glass. (4112, Pb-1567)
e + p 1570	Guan Yin, Song-Ming, or 12th-17th cent. Slightly turbid colorless glass, some evidence of light weathering or crizzling. ROM 922.20.80A. (5802, Pb-1570)

Non-Glasses and Non-Chinese

* + 08	Fragment of glazed ceramic, 18th cent., made at Jingdezhen. Sample is of bright yellow glaze. Collection of E.T. Hall, Oxford. (Pb-1008)
* s + 56	Large eye bead, probably Near Eastern 600-250 B.C. Turbid light blue glass with large knobbed black and white eyes and numerous other applied blobs, heavily weathered. Sample is of weathered yellow opaque glass knob. CMG unaccessioned. (3337, Pb-1556)
+ 72	Metal buckle (probably bronze), gilded, with blue opaque glass inlays, 8-17th cent. A.D. Sample is of black metallic corrosion products. ROM 950.77.3. (5807, Pb-1572)

e	+	97	Faceted bar of "Chinese purple", Han (?). Sintered molded bar of lavender color, slightly weathered. Electron microprobe analysis shows presence of some lead. ROM 981.5.6. (5799, Pb-1597)
	+	1599	Glazed terra cotta figure of an animal, date uncertain, possibly Chinese; purchased in Ghazni. Sample is of weathered green glaze. R.H.B. collection. (Pb-1599)
	+	2000	Faceted bar of "Chinese purple", Han (?). Sintered molded bar of light purple color. ROM sample no. 4, Zycherman and FitzHugh sample e. (Pb-2000)
	+	2001	Faceted bar of "Chinese purple", Han (?), possibly from Jincun, near Luoyang. Sintered molded bar of dark purple color. ROM sample no. 8, Zycherman and FitzHugh sample f. (Pb-2001)
	+	2010	Ceramic tile, Tanzhe Monastery, date uncertain. Blue transparent glaze. Collected by R.H.B., 9/12/84.
	+	2011	Ceramic tile, Tanzhe Monastery, date uncertain. Yellow transparent glaze. Collected by R.H.B., 9/12/84.
* s	+	2076	Bangle, date uncertain, possibly 12-14th cent. Excavated at Kota Cina, Sumatra, by E. Edwards McKinnon. (See note 9.) White opaque glass, lightly weathered. Opacified with CaF_2. Excavation no. 51.T4.EO.60, 2/22/76. (2926, Pb-2076) $(K_2O+CaO):PbO:SiO_2$; PbO = 47.6%.
* s	+	2077	Large barrel-shaped bead, same date, source, and lot as 2076. Yellowish turbid glass, moderately weathered. (2927, Pb-2077) $(K_2O+Na_2O):PbO:SiO_2$; PbO = 37.0%.
* s	+	2078	Large spherical bead, same date, source, and lot as 2076. Yellow opaque glass, unweathered. (2928, Pb-2078) $(K_2O?):PbO:SiO_2$; PbO = 53.2%.

Ores and Minerals

	+	1804	Bournonite (PbCuSbS), Guangdong Province. Courtesy Paul E. Desautels, National Museum of Natural History, Washington; NMNH 90614. (Pb-1804)
	+	1805	Galena, Shuikoushan, Hunan Province, Courtesy John E. Dayton. (Pb-1805)
	+	1806	Galena, Jiangxi Province. Courtesy Liu Yongquan, Geological Museum, Beijing. (Pb-1806)
	+	1807	Galena, Liaoning Province. Courtesy Liu Yongquan. (Pb-1807)
	+	1808	Galena, Yunnan Province. Courtesy Liu Yongquan. (Pb-1808)
	+	1809	Galena, Hunan Province. Courtesy Liu Yongquan. (Pb-1809)
s		1598	Remingtonite (?) (hydrated cobalt carbonate), prob. Yimen Mine in Yunnan or Jinchuan Mine in Gansu. Courtesy Hu Xigeng, Central-South Institute of Mining and Metallurgy, Changsha. Emission spectrographic analysis shows ~2% Pb, ~1% Ba and major Mn, Fe, and Al. (4780, Pb-1598)
* s	+	2050	Hokutolite, a lead-containing barite, Hokutu, Taiwan. Soft, dense, tan-colored polycrystalline material. PbO = 6.04%; BaO = 52.0%. (4781)

ADDENDUM TO CHAPTER 5

ROBERT H. BRILL, SHI MEIGUANG, EMILE C. JOEL, AND ROBERT D. VOCKE

The Corning Museum of Glass, Corning, N.Y.; China Building Materials Academy, Beijing; The Smithsonian Institution, Washington, D.C.; The National Institute of Science and Technology, Gaithersburg, Md.

We present here lead isotope analyses of several additional samples analyzed while this book was in proof. They are samples nos. 2002-04, 2012-13, 2020-21, 2028-29, 2082, and 3400. A concordance of these sample numbers and their corresponding chemical analysis numbers is given below. The data are reported in Table 1 and plotted in Figure 1.

The additional samples include five early glasses, two of which are Shi Meiguang's Xuzhou samples, nos. 2012 and 2013. (See Addendum to Chapter 4.) Both are lead-barium glasses but, although found in the same tomb, their chemical compositions differ somewhat. This difference carries over into their lead isotope ratios which vary dramatically, one being quite high, the other quite low. It is tempting to speculate that the two leads came from widely separated sources — and that may well be true. However, one has to keep in mind that there are mining regions in China where lead-bearing ores of high and low ratios occur relatively close together. For example, at Huize in Yunnan Province, ores with $^{207}Pb/^{206}Pb$ ratios of 0.860 and 0.762 are both found.[1] The numerical difference between the ratios of these two ores is about the same as that between the leads in the two Xuzhou glasses.

Sample no. 2021 came from the eyebead beaker discussed at some length in the Addendum to Chapter 4. As was stated there, the lead isotope ratios for the blue glass in the beaker and that in the Royal Ontario Museum fragments are identical, within the estimated experimental error. We believe the fragments and reconstructed beaker came from a single parent object.

The two other early glasses analyzed are nos. 2082 and 2083, the two Singer plaques also discussed in the Addendum to Chapter 4. As can be seen in Table 1, the isotope ratios determined for these two glasses are in close agreement,[2] and they can reasonably be assumed to have come from the same mining region as one another. Because their chemical analyses also match so closely, the glasses were probably made in the same place at about the same time.

Isotopically, the two plaques are a fairly close match to a previously established group of four glasses described in Chapter 4 (nos. 05/06, 09, 80, and 83). We see them all as having a common origin. It is also worth noting that all these glasses have a similar colorless but turbid appearance, probably imitating jade.

Samples nos. 2002 and 2003 are from sintered bars of Chinese purple/blue. In appearance, they are very much like the two bars discussed in Chapter 5 (nos. 2000 and 2001). The lead in no. 2003 is virtually identical to the lead in those same bars, and it would be fair to assume that all three were made in the same place. They have the same type of lead as one of the high-range ratio groups of barium-containing glasses. The other bar analyzed recently, no. 2002, has lower ratios, not matching, but also not very different from three early Chinese glasses (nos. 07, 53, and 94). Still another example of Chinese blue — the brilliant blue, powdery coating on the Singer terra cotta bead, no. 2004 — is, again, a very close match to the three Chinese purple bars just mentioned. Apparently Chinese blue pigment (the tetrasilicate) and sintered bars of Chinese purple (the disilicate) were made in the same region, most likely in the same workshop. It is noteworthy, too, (although not at all unexpected) that these products of what we regard as a rather sophisticated pyrotechnology appear to have been made in the same region as some of the early Chinese glass — itself the product of a noteably sophisticated pyrotechnology.[3]

No. 3400 is an extremely important sample. It comes from a fragment of one of the blue faience beads found in Baoji or Fufeng County, west of Xian. These beads are from Western Zhou tombs, several of which in that region have been dated to the 9th-8th centuries B.C. Once thought to have been made of glass, the beads have now been demonstrated to be made of faience.[4] The trace of lead in the blue vitreous phase of a chip of this fragment proved to have low isotope ratios, but not as low as the group of glasses just discussed. Instead, it falls in the middle of a wide gap separating the major isotope fields established for early Chinese glasses. (See Figure 1.) It does not match any of the Chinese glasses of any periods we have analyzed so far. Neither does it match the ratios reported by Wang Shixiong (see Chapter 15) for a faience bead of similar origin. The latter is a good match for four Chinese glasses with mid-range values. They are nos. 51, 57, 64, and 82 — all Han glasses — plotted in Figure 2 of Chapter 5.

The interpretation of the data for no. 3400 is clearly a complicated matter and any inferences drawn must, for the present, be tentative. We will not discuss them here but instead refer the reader to a publication where the results are discussed.[5] The main point to be emphasized here is that because of the archaeological importance of these early faience beads, every effort should be made to learn as much as we can about as many of them as we can.

No. 2020 is a light blue opaque bowl acquired by the Metropolitan Museum of Art in Egypt many years ago. Although originally thought to be Islamic, it is now catalogued as Chinese. An Jiayao, upon examining a photograph, told one of the authors (R.H.B.) that she believes the bowl is of the Yuan Dynasty (1271-1368). We had previously wondered if it might have come from Fustat, in which case somewhat earlier dates (perhaps from the 9th to 13th centuries) would have been possible. James Watt has observed that Fustat has yielded Islamic pottery imitating in a curious way, certain Yuan Dynasty pottery made by coiling clay on the top of an already fired plate and refiring the assembly after the application of a heavy glaze. The imitations made at Fustat even had Chinese symbols painted on them. He thought there might possibly be some connection between the glass bowl and that Fustat pottery which, in turn, would be consistent with a Yuan Dynasty date for the glass. The glass contains 13.89% PbO, which, isotopically, is very similar to that in a small Tang Dynasty Buddha (no. 1588), and which lies intermediate between two groups of somewhat later Chinese glasses. (See Figure 3 in Chapter 5.)

The final sample is no. 2082, lead extracted from a "false gold glass" bead acquired a few years ago from islanders on Botago Tobago in the Philippines. The bead is made of yellowish glass (actually pale amber) sandwiching silver foil, giving it the appearance of gold glass. It is undated but heavily pitted and shows much wear.

Dezso Benedek, who acquired the bead, says it is similar to beads worn by "Formosan mountain aborigines." Years ago, our good friend Dr. van der Sleen published a gold glass bead that resembles the bead in question.[6] It was from Flores, and it had been described by H.C. Beck[7] as Romano-Egyptian. Beck probably saw it as being similar to certain Hellenistic beads, such as those excavated at the 3rd-2nd century B.C. bead factory at Rhodes.[8] Another gold glass specimen resembling no. 2082 is illustrated in a Japanese publication,[9] but we cannot tell whether it and the other example are truly gold glass, or if they are the "false gold glass" made of an amber glass sandwiching silver foil.

The glass in no. 2082 has a K_2O:PbO:SiO_2 composition, containing 41.9% PbO; thus it is certainly not of Hellenistic origin.[10] Three other glasses we have studied (nos. 1560, 1563, and 1570) almost certainly of Chinese origin, have the same type of chemical composition and also have similar lead isotope ratios. They are described in Chapters 4 and 5. They range in date from the 17th to the early 20th centuries. On the strength of this, it seems quite possible that the bead Benedek acquired in the Philippines was made in China.

It would be interesting to explore this matter further, especially if additional samples of the "false gold glass" variety could be made available for analysis. This, however, is just one aspect of a larger picture. The whole field of lead isotope analysis of Chinese and other East Asian glasses is so extraordinarily promising, that we hope it will be possible to maintain any momentum this modest start has provided.

Notes

1. This volume, Chapter 16.
2. Conservative estimates of the standard deviations for the $^{207}Pb/^{206}Pb$ and $^{208}Pb/^{206}Pb$ ratios are about 0.1% of the nominal values. These correspond to 2 sigma limits of ±0.004 and ±0.002, respectively, for the two ratios. The two glasses mentioned differ by just about these values.
3. For further information on these compounds, see this volume, Chapters 4 and 5.
4. See for example this volume, Chapters 14 and 15, and both Introductions. See also: R.H. Brill, S.S.C. Tong, and Zhang Fukang, "The Chemical Compositions of a Faience Bead from China," *Journal of Glass Studies,* vol. 31, 1989, pp. 11-15; and R.H. Brill, R.D. Vocke, Wang Shixiong, and Zhang Fukang, "A Note on Lead-Isotope Analyses of Faience Beads from China," *Journal of Glass Studies,* vol. 33, 1991.
5. See R.H. Brill, R.D. Vocke, Wang, and Zhang cited in note 4.
6. W.G.N. van der Sleen, *A Handbook on Beads,* Liege, The International Association for the History of Glass, 1973, p. 98.
7. H.C. Beck, *Archaeologia,* vol. 77, (1928).
8. Gl. Davidson Weinberg, "Glass Manufacture in Hellenistic Rhodes," *Archailogikon Deltion,* vol. 24, 1969, pp. 143-151 and plates 76-88.
9. *Ancient Glass from the Orient, February 7 - March 12, 1978,* Tokyo National Museum, 1978, p. 144, no. 534. The bead was excavated in a Kofun Period (late 5th century) tomb in Nara. It is described as containing gold leaf.
10. We believe that no potash-based glasses were made in the West until Mediaeval times; certainly they were not made there in anything like significant quantities during Hellenistic or Roman times. For example, our unpublished analyses of the glass from the Rhodes factory show that glass to have been a natron-type soda lime. However, K_2O:SiO_2 glasses were made in China or Southeast Asia about as early as the Hellenistic Period. See Shi Meiguang, He Ouli, and Zhou Fuzhen, "Investigations of Some Chinese Potash Glasses Excavated from Tombs of the Han Dynasty," *Journal of the Chinese Silicate Society,* vol. 14, 1986. For examples of early K_2O:SiO_2 and K_2O:CaO:SiO_2 glasses from Korea, Ban Don Ta Phet, and Ban Chieng, see R.H. Brill, "Scientific Investigations of Ancient Asian Glass," in the proceedings of the Nara Symposium '91, Unesco Maritime Route of Silk Roads, March 7, 1991 (in press).

Table 1.

Isotope Ratios of Lead in Chinese Glasses and Related Materials

Sample number	*NBS/NIST number*	$^{208}Pb/^{206}Pb$	$^{207}Pb/^{206}Pb$	$^{204}Pb/^{206}Pb$
Early Period Glasses				
2012	2385	2.1764	0.86751	0.05583
2013	2386	1.9547	0.74290	0.04647
2021	2372	2.1901	0.89019	0.05747
2028	2373	1.9185	0.71719	0.04456
2029	2374	1.9224	0.71918	0.04458
Early Period Faience				
3400	3333	1.9621	0.79111	0.04938
Early Period Chinese Blue/Purple				
2002	3.337	2.0851	0.80802	0.05131
2003	3.335	2.1873	0.88724	0.05720
2004	2371	2.1854	0.88786	0.05722
Middle Period Glasses				
2020	2375	2.1314	0.86493	0.05544
Other				
2082	3-331	2.1179	0.85653	0.05478

Figure 1. *Sample no. 2020. (Same as anal. no. 4120.)*

Figure 2. *Sample no. 2082.*

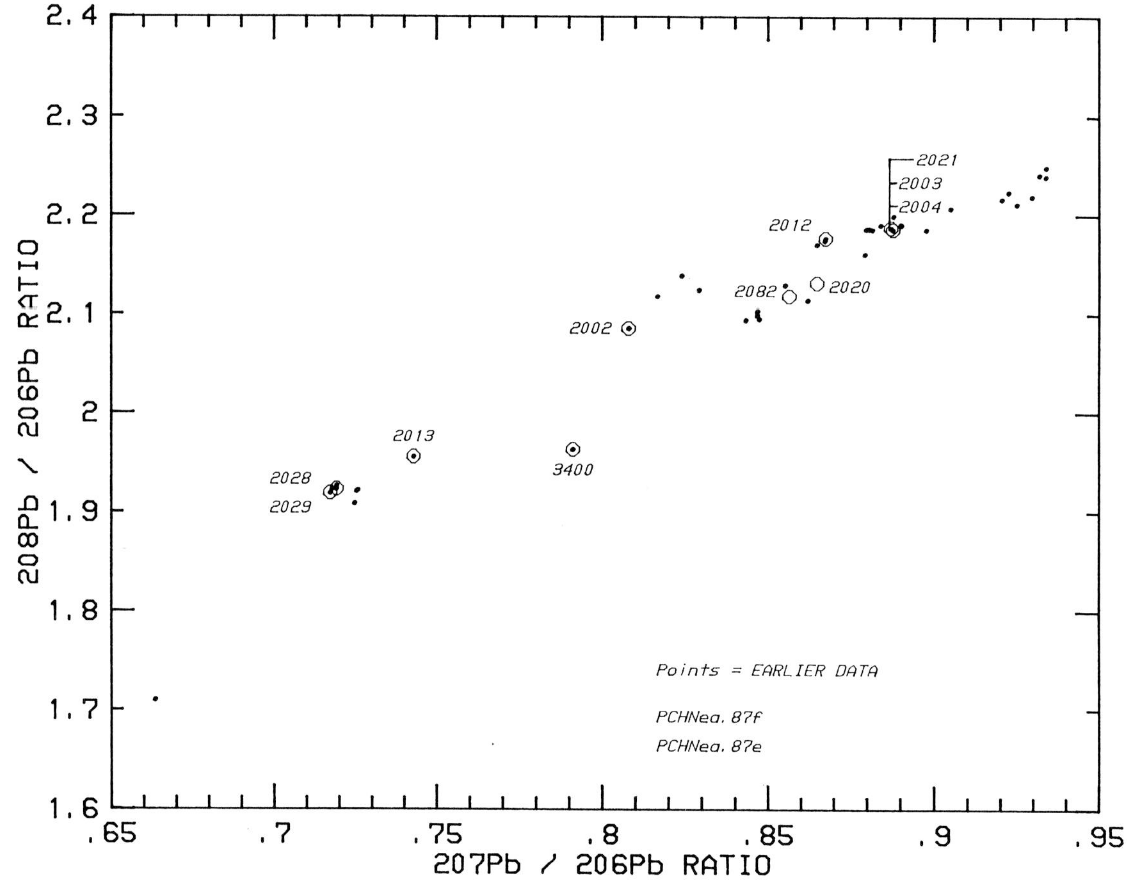

Figure 3. *Isotope ratios for lead extracted from additional samples. For comparison, data from figures in Chapter 5 are replotted here as dots.*

CATALOGUE OF SAMPLES

This concordance gives the lead isotope numbers of the samples discussed in this addendum, along with their corresponding CMG chemical analytical numbers. If they do not appear here, detailed sample descriptions can be found in the catalogues at the end of Chapter 4 or Addendum to Chapter 4. For convenience, we have dropped our customary Pb- prefix to the lead sample numbers here and throughout the text.

2002		Octagonal bar of "Chinese purple". Light purple color with some blue particles. Östasiatiska Museet, Stockholm, no. K4069. Zycherman and FitzHugh sample h.
2003		Long bar of "Chinese purple". Purple with colorless particles. As above, K4070.
2004	CMG 5894	Chinese blue pigment on terra cotta bead. (P. Singer.)
2012	CMG 3382	Xuzhou cup. (Shi Meiguang.)
2013	CMG 3383	Xuzhou animal. (Shi Meiguang.)
2020	CMG 4120	MMA lt. blue opq. bowl. (Christine Lilyquist, James C.Y. Watt.)
2021	CMG 3345	"Eyebead" cup. (Eskenazi, Ltd.)
2028	CMG 5892	Plaque from "glass burial suit". (P. Singer.)
2029	CMG 5893	As above, from a different group. (P. Singer.)
2082	CMG 5536	Medium, barrel-shaped bead, Botago Tobago Island, or nearby in the Philippines; undated. P. yellowish glass with silver (?) foil sandwiched in, giving appearance of gold glass; pitted, with wear. The bead is said to be similar to those worn by "Formosan mountain aborigines." From Dezso Benedek. AK_2O:PbO:SiO_2 glass, with 41.9% PbO.
3400	CMG 5895	Fragment of a faience bead, Shaanxi, 9th-8th c. B.C. Blue glassy phase, extending through body, contains 70.6% SiO_2, 15.0% K_2O, 8.16% CuO, 2.29% Na_2O, 2.85% Al_2O_3, 0.30% MgO, 0.56% Fe_2O_3, and 0.0X% Pb. See R. H. Brill, S. S. C. Tong, and Zhang Fukang, "The Chemical Composition of a Faience Bead from China," *Journal of Glass Studies*, 1989, v. 31, pp. 11-15.

CHAPTER 6

SIMILARITIES BETWEEN ANCIENT CHINESE GLASSES AND GLASSES EXCAVATED IN JAPANESE TOMBS

KAZUO YAMASAKI AND MASAYO MUROZUMI

Nagoya University, Chikusa, Nagoya;
Muroran Institute of Technology,
Hokkaido

THE early cultural history of Japan is divided into the Jomon and Yayoi Periods, both named after types of pottery. No glass from the Jomon Period (earlier than about the 3rd century B.C.) has been found. However, glass beads have been uncovered at several sites of the Yayoi Period (ca. 3rd century B.C.-3rd century A.D.).

In Fukuoka Prefecture, located in the northern part of Kyushu, there are several important archaeological sites of the Yayoi Period. These include Sugu-Okamoto near Fukuoka City and Tateiwa, Iizuka City, which is about 30 km northeast of Fukuoka City. At these sites, which date to about the 1st century B.C., large pottery jars used as coffins were excavated. Glass beads and other artifacts such as bronze mirrors, bronze weapons, and shell bracelets were often found inside coffins. After the Yayoi Period came the protohistoric Tumulus Period (ca. 3rd-7th centuries A.D.) during which huge mounds were constructed for the burial of the dead. Glass beads have been found in most of these tombs. The majority of these beads are round or comma-shaped (*magatama*), and they appear to have been used as ornaments. The comma-shaped beads are characteristic of Japan and southern Korea and were certainly made in these areas (Table 1, Fig. 1). The other beads seem to have been made elsewhere.

The end of the Tumulus Period overlaps the historic Asuka-Nara Period (552-793 A.D.) without clear demarcation. With the coming of Buddhism to Japan in the 6th century and during the Asuka-Nara Period, the making of glass beads in Japan reached a high level. These beads are mostly lead glasses made from Japanese lead ores.

In 1934, Seligman and Beck first noted the unusually large amounts of PbO and BaO in some pre-Han glass beads.[1] Since 1950, one of the authors (K.Y.) has conducted chemical studies of glass beads excavated in Japan, especially those found in the sites of the Yayoi Period in the northern part of Kyushu such as Sugu-Okamoto and Tateiwa. Most of these beads are made of glass rich in PbO and BaO. Some of the results of those studies and of Chinese glasses are presented in Table 2.[2]

The chemical compositions of the glasses excavated in the northern part of Kyushu are closely related to those of two Chinese ritual disks from the Warring States Period (475-221 B.C.) and also to the composition of a glass cicada studied by Seligman and others.[3] Because such high concentrations of BaO had previously been found only in pre-Han glasses, it is highly probable that these beads (nos. 3-6) were made either in Japan from imported glass lumps, or in China during pre-Han times. A sandstone mold used for casting comma-shaped beads and a glass bead with a high BaO content which fits in this mold were found in different sites of the same period in Fukuoka Prefecture.[4] This indicates that trade in glass might have been conducted between Japan and China during the Yayoi Period.

In 1983, another comma-shaped bead was found in Kumamoto Prefecture (see no. 6 in Table 2).[5] The site is situated about 100 km south of Fukuoka City, and its location indicates an expansion of the use of PbO-BaO glasses into the southern and inner parts of Kyushu. This is important from an archaeological point of view.

The lead isotope ratios of the glasses are given in Table 3. They are also shown (Fig. 2) with lead isotope ratios of a recently found comma-shaped bead.[6]

The lead isotope ratios of glasses nos. 1-5 were determined at the National Bureau of Standards, U.S.A., and reported in 1979,[7] while those of the recently found bead, no. 6, were determined in the Muroran Institute of Technology, Hokkaido.[8] These ratios are not identical, but are in the same range as those in Figure 2. The isotope ratios of bead no. 4 are now being redetermined because the weathered portion of this bead showed different values. This bead has remarkably low isotope ratios.

As shown in Figure 2, the leads in the five glass beads with high PbO and BaO contents (nos. 1, 2, 3, 5, and 6) are very likely to have come from the same mining region, but not from the same mine. In order to identify a possible source of manufacture for the early BaO-rich glasses, lead isotope data for Chinese galena ores are required. {Editor's Note: See Chapter 16.} It is very difficult to obtain data for Chinese galena ores, and we have so far

Table 1. Chronology of China and Japan

CHINA	JAPAN
Western Zhou 770 B.C.	Jomon Period
Eastern Zhou Spring & Autumn 475 B.C.	
Warring States Period 221 B.C.	ca. 300 B.C.
Qin 206 B.C.	
Western Han 24 A.D.	Yayoi Period
Eastern Han 220	
Three Kingdoms	ca. 300 A.D.
Jin 420	Tumulus Period
Southern and Northern Dynasties 581	552
Sui 618	Asuka 645
Tang 907	Nara 794
Five Dynasties 979	Heian
Northern Song 1127	1185

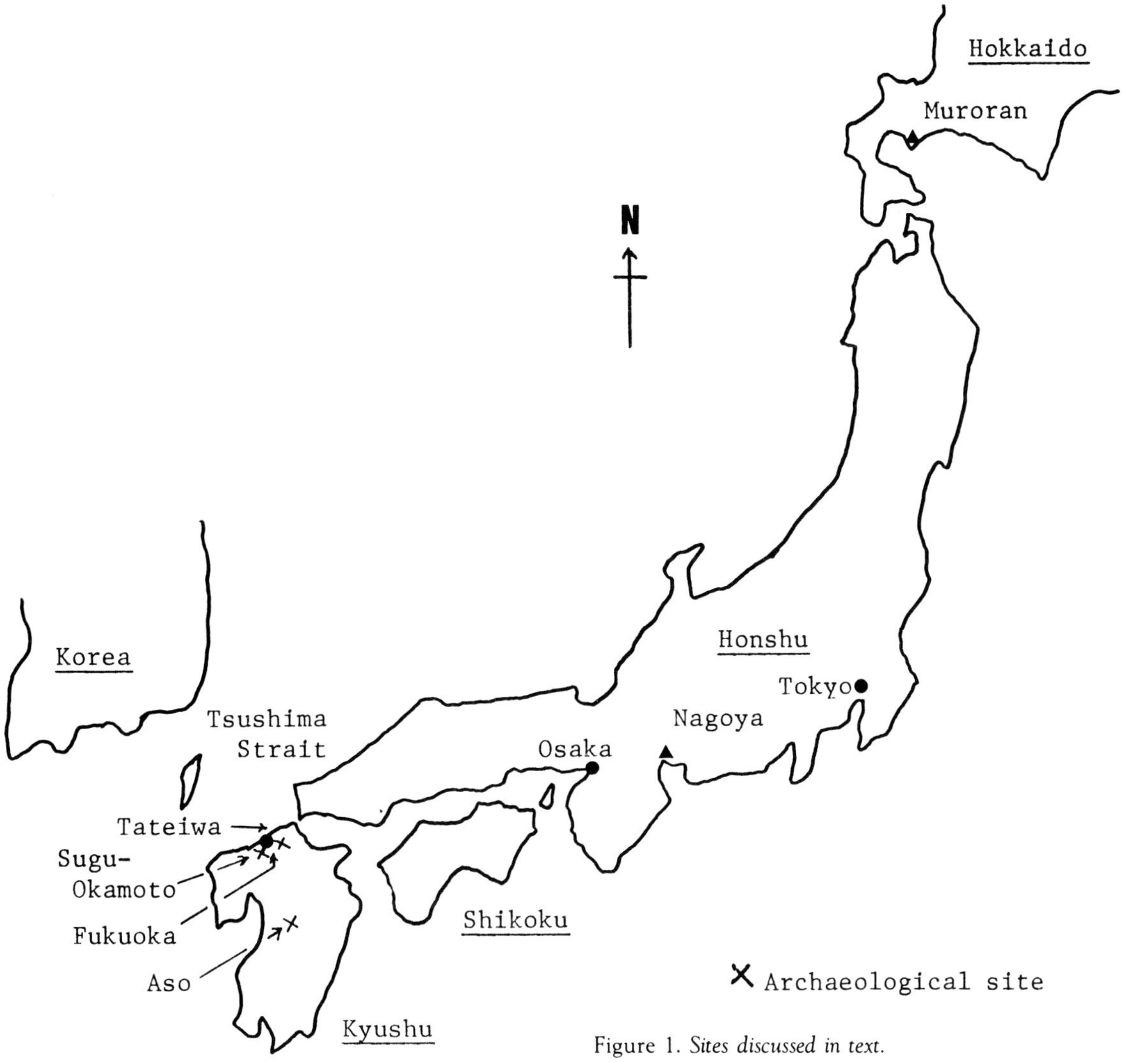

Figure 1. *Sites discussed in text.*

studied only three samples. Two of these are from Liaoning Province and one is from Hunan Province. Their isotope values, however, do not agree with those of the glasses with high levels of BaO.

Because the galena samples are few, we have compared data on the glasses with data on some Chinese bronze mirrors and other objects containing lead in an effort to determine the place of production of the BaO-rich glasses. The Western Han and Warring States mirrors showed rather high isotope ratios, but they are not as high as those of the PbO-BaO glasses. However, two halberd samples (the type without sockets) in the collection of the Department of Archaeology of The University of Tokyo showed very low isotope ratios, which are in the same range as glass no. 4 (Fig. 2), and glasses nos. 1005 (a ritual disk), and 1009 (a *chape*) (See ref. 3). The isotope data of these two halberds are shown in Table 4.[8]

Unfortunately the provenances of these two halberds are unknown, but presumably they belong to the Shang Period (16th-11th centuries

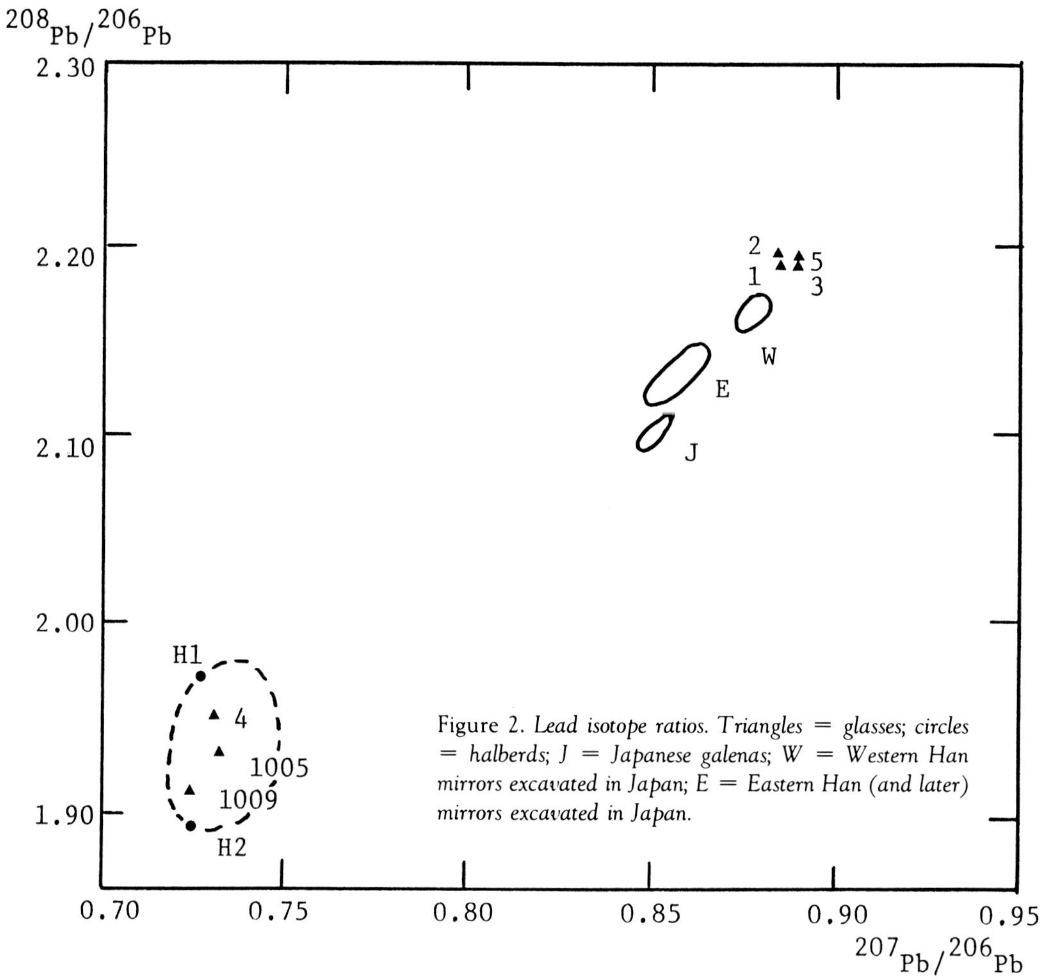

Figure 2. *Lead isotope ratios. Triangles = glasses; circles = halberds; J = Japanese galenas; W = Western Han mirrors excavated in Japan; E = Eastern Han (and later) mirrors excavated in Japan.*

B.C.). Thus we have now obtained information that Shang bronzes and some pre-Han glasses seem to have been made using lead from mines with similar geological ages, possibly from mines in the same region as one another.

From the viewpoint of Japanese archaeology, the finding of a high PbO-high BaO glass in the Kumamoto Prefecture in central Kyushu is important. The Fukuoka Prefecture along the Tsushima Straits has been the important gateway to Korea and China for more than two thousand years, and it is not surprising to find glasses there which are related to the pre-Han periods. On the other hand, the site in Aso-cho, Kumamoto Prefecture, where a comma-shaped bead with a PbO-BaO composition was recently excavated, is situated in the inner part of Kyushu far from the Tsushima Straits. This illustrates the expansion of the area influenced by overseas culture into the inner and central part of Kyushu. Moreover, it seems reasonable to assume that this site with its mid 3rd-century A.D. date was related to a powerful family which used ornaments made from imported materials.

Experimental. The recently found comma-shaped glass bead with high levels of PbO and BaO was analyzed by the so-called "one drop" atomic absorption spectrometry technique.[9] Because the sample was small (0.23 mg), only BaO, Na_2O,

K_2O, PbO, and CuO were detected. The lead isotope ratios of glasses and mirrors were determined on a Hitachi RMU-6 mass spectrometer combined with a microcomputer. The data were normalized by analyzing the standard sample NBS-SRM 981.

Acknowledgments. The authors are grateful to many individuals who provided the glass and bronze samples used for this research. Prof. C. Iida, Nagoya Institute of Technology, and his co-workers helped in the chemical analyses. Mr. S. Nakamura and many students in the Muroran Institute of Technology helped in the preparation of the samples and the mass spectrometry. Part of the expenses was defrayed by a Grant-in-Aid for Scientific Research of the Ministry of Education, Science, and Culture. We also thank Mr. Shi Meiguang (China Building Materials Academy, Beijing), Dr. R. H. Brill (The Corning Museum of Glass), Mr. Zhang Guoxin (Institute of Geochemistry, Guiyang), Mr. W. T. Chase (Freer Gallery, Washington) and Dr. H. Mabuchi (Tokyo National Research Institute Cultural Properties) for providing valuable information and helpful discussions.

Post-Symposium Notes

1. At the Symposium, Shi, He, Wu, and Zhou reported chemical analyses of thirteen glass samples, four of which were high PbO-BaO glasses.[10] Their compositions were similar to those of our glasses nos. 1-6 (Table 3). Brill, Tong, and Dohrenwend,[11] also at the Symposium, reported eighteen analyses of the high PbO-BaO type.

2. At the Symposium, Barnes, Brill, and Deal reported isotope data on Chinese glasses of various periods.[12] Some glasses showed very high and others very low isotope ratios. The low values are close to those of glass No. 4 (Fig. 2). Thus the number of PbO-BaO glasses showing very low isotopic compositions has increased.

3. In 1980 Chen, Mao, and Zhu reported lead isotope compositions of more than 120 lead ores from various parts of China.[13] Among these ores one from the Yongshan Jinsha mine, Yunnan Province, showed very low values: $^{207}Pb/^{206}Pb = 0.7624$, whereas others showed $^{207}Pb/^{206}Pb = 0.82\text{-}0.88$. In 1984 Yin, Yu, Zhang, and Lu reported lead isotope compositions of ores of Chinese mines.[14] One of them, from the Kaobanhe mine in Hebei showed very high values: 0.9184-0.9987. Judging from the above data a trend seems to exist that the ores of the northern parts in China show higher ratios while those in southern parts show lower ratios.

4. Jin Zhengyao read a paper at the Third International Conference on the History of Science concerning the sources of ores for the bronzes produced in the Central Plain during the late Shang period. He reported that ores of the Yongshan Jinsha mine in Yunnan Province showed very low isotope ratios, while those of the Chaihe mine in Liaoning had very high values. He did not give detailed data, but judging from his plot of the data, those of glass no. 4 and the two halberds fall in the region of the Yongshan Jinsha mine.

5. Recently Mabuchi determined lead isotope ratios of some Chinese, Korean, and Japanese lead ores.[15] Among them the lead ore of Huize mine in Yunnan showed very low values, and those of the Chaihe mine in Liaoning showed very high values. Both Yongshan Jinsha reported by Jin and Huize are situated in the northern part of Yunnan, and it seems highly probable that high PbO-high BaO glasses having very low isotope ratios were made with lead from that mining region.

Catalogue of Samples

1. Fragment of Chinese glass ritual *bi* disk. Believed to have been excavated in Changsha, Hunan, Warring States Period. Dark green, heavily weathered. Donated by S. Umehara. Same as no. 1520 in note 2, Brill, *et al.*

2. Fragment of Chinese glass ritual *bi* disk. Warring States Period. From the Eguchi Collection. Dark green, heavily weathered. Donated by S. Umehara. Same as no. 1521 in note 2, Brill, *et al.*

3. Cylindrical tubular glass bead. Excavated at Sugu-Okamoto, in pottery coffin, Kasuga City, Fukuoka Prefecture. 1st century, B.C. Length ca. 2 cm, diam. 5 mm. Dark green, heavily weathered. Donated by S. Umehara. Same as no. 1523 in note 2, Brill, *et al.*

4. Rod or pestle-shaped glass. Excavated at Tateiwa, in a pottery coffin No. 28, Iizuka City, Fukuoka Prefecture. 1st century B.C. (latter half)-1st century A.D. (first half). Length ca. 3 cm, width 1.5 cm. Dark green, heavily weathered. Donated by T. Okazaki. Same as no. 1524 in note 2, Brill, *et al.* The isotope data were determined for the corrosion products, but it did not give consistent results. Therefore, the determination was repeated on glass from underneath the corrosion layer.

5. Cylindrical tubular glass bead. Excavated at Uki-Kunden, in pottery coffin No. 106, Karatsu City, Saga Prefecture (previously incorrectly reported as coffin No. 10). Late Yayoi Period. Length 8 mm, width 7 mm. Dark green, heavily weathered. Donated by T. Okazaki. Same as no. 1526 in note 2, Brill, *et al.*

6. Comma-shaped glass bead (or *magatama*). Excavated at Shimoyamanishi, Otohime, Aso-cho, Kumamoto Prefecture. Late Yayoi Period. Dark green, weathered. The details will be reported later by the archaeologists of the Kumamoto Prefectural Office.

Notes

1. C. G. Seligman and H. C. Beck, *Nature,* 133, 1934, p. 982; C. G. Seligman, P. D. Ritchie, and H. C. Beck, *Nature,* 138, 1936, p. 721; C. G. Seligman and H. C. Beck, *Bulletin of the Museum of Far Eastern Antiquities,* no. 10, 1938, p. 1.
2. K. Yamasaki, *Proceedings of the Tenth International Congress on Glass* (Kyoto), no. 9, 1974, pp. 15-20; R. H. Brill, K. Yamasaki, I. L. Barnes, K. J. R. Rosman, and M. Diaz, *Ars Orientalis,* 11, 1979, p. 87.
3. See note 1.
4. S. Umehara, *Shirin,* 43, 1960, p. 1. (In Japanese.)
5. The archaeological excavation of the site, Shimoyamanishi, Otohime, Aso-cho, Kumamoto Prefecture, was carried out by a group from the Kumamoto Prefectural Office. A detailed report will be published in the near future.
6. See note 2.
7. *Ibid.*
8. Unpublished data of K. Yamasaki and M. Murozumi.
9. T. Uchida, C. Iida, and I. Kojima, *Anal. Chim. Acta,* 113, 1980, p. 361; T. Uchida, I. Kojima, and C. Iida, *Anal. Chim. Acta,* 116, 1980, p. 205; I. Kojima and C. Iida, *Analyst,* 107, 1982, p. 1000; I. Kojima, C. Iida, and K. Yamasaki, *Bunseki Kagaku,* 31, 1982, p. E167. (In English.)

Post-Symposium Notes

10. This volume, Chapter 3.
11. This volume, Chapter 4.
12. This volume, Chapter 5.
13. Chen Yuwei, Mao Cunxiao, and Zhu Bingquan, *Geochimica,* 1980, p. 215. (In Chinese.) See also the English summary of their results included in this book as Chapter 16.
14. Yin Hanhui, Yu Cimei, Zhang Guoxin, and Lu Jialan, *Scientia Sinica,* Ser. B, 27, p. 852, 1984. (In English.)
15. Personal communication.

Table 2

Chemical Analyses of Some Chinese and Japanese Glasses

	China, Ritual disk 1 (1520)	China Ritual disk 2 (1521)	Sugu-Okamoto, Tubular bead 3 (1523)	Tateiwa, Rod-like bead 4 (1524)	Uki-Kunden, Tubular bead 5 (1526)	Aso, *Magatama* 6	China, Cicada (Seligman *et al.*)[1] 7
SiO_2*	36	36	38	37	38.9		34.42
Al_2O_3	0.36	0.13	0.35	0.52	0.56		0.76
Fe_2O_3	0.13	0.07	0.29	0.14	0.30		0.16
CaO	1.6	0.01	1.1	0.50	2.83		0.12
BaO*	14	13	14	16	7.59	11.1	12.58
MgO	0.16	0.08	0.15	0.20	0.95		0.35
Na_2O	1.65	1.96	3.90	2.65	4.68	3.91	4.32
K_2O	0.26	0.17	0.19	0.50	0.21	0.09	1.02
PbO	46.1	48.5	38.5	40.8	43.5	38.0	43.20
CuO	0.88	0.84	0.78	0.76	0.42	0.81	
Method	aa	aa	aa	aa	w	aa	w

* SiO_2 and BaO for samples 1-4 were estimated by emission spectrography.

aa Atomic absorption.

w Wet chemistry.

Table 3

Lead Isotope Data for High PbO-High BaO Glasses

Sample	$^{207}Pb/^{206}Pb$	$^{208}Pb/^{206}Pb$	$^{206}Pb/^{204}Pb$
1. China, ritual disk	0.8846	2.1930	17.55*
2. China, ritual disk	0.8840	2.1939	17.57*
3. Sugu-Okamoto, tubular bead	0.8881	2.1933	17.45*
4. Tateiwa, tubular bead	0.7301 (0.05)	1.944 (0.04)	22.17(0.15)**
5. Uki-Kunden, tubular bead	0.8898	2.1939	17.42*
6. Aso, comma-shaped bead	0.8824(0.09)	2.181(0.16)	17.62(0.19)**

* Data for nos. 1, 2, 3 & 5 from ref. 3, Brill *et al.* These samples correspond to nos.1520, 1521, 1523, and 1526 in ref. 3.

** No. 4 (no. 1524 in ref. 3) was re-analyzed using the glass body. The figures in parentheses indicate the coefficient of variation.

Table 4

Lead Isotope Data of Halberds

	$^{207}Pb/^{206}Pb$	$^{208}Pb/^{206}Pb$	$^{206}Pb/^{204}Pb$
1. Halberd 1 (sample B97) (Univ. Tokyo No. C-70)	0.7296(0.3)	1.972(0.3)	21.73(0.5)
2. Halberd 2 (sample B100) (Univ. Tokyo No. C-76)	0.7227(0.1)	1.883(0.1)	21.77(0.3)

CHAPTER 7

CHINESE GLASS TECHNOLOGY IN BOSHAN AROUND THE 14TH CENTURY

YI JIALIANG AND TU SHUJIN

The Glass & Enamel Research Institute,
Ministry of Light Industry,
Shanghai

IN December, 1982, the site of an ancient glass factory was discovered in the city of Zibo in Shandong Province. It was located on the northern part of Da Street by the bank of the Shaofu River, a place which formerly belonged to the old town area of Boshan.

About 20 furnace pits were excavated there. Examples of glass beads, hair clasps, and ornaments, together with the remains of clay pots and waste glass, were gathered by the authors from one of the furnace pits located 1.6-1.8 m underground. Two cultural strata of later date (with an alluvial layer between them) were located above the level of the excavated furnace pits. The latter is believed to be the site of an ancient glass factory of a considerable scale of production dating back to the time between the late Yuan Dynasty (1271-1368) and early Ming Dynasty (1368-1644).[1]

Several unearthed samples of glasses, melting pots, and pot covers were examined by various methods. Their chemical analyses are reported in Table 1.

Glass Samples

Pieces of well-melted blue opal glass remaining in the excavated furnace pots were selected as samples for testing. The original surfaces of the unearthed glass were covered with a thick, white, weathered layer resulting from the long period of burial in the earth.

Infrared spectra and x-ray diffraction show that the main phase is a glassy substance. The chief absorption peak of the IR spectra is near 1000 cm^{-1}, which corresponds to the bond-stretching vibrations of the Si-O linkage in glass. In addition to the diffuse peak of the glassy phase, sharp peaks corresponding to calcium fluoride (CaF_2) crystals are found in the XRD pattern.

A sharp contrast between the bright crystals and the dark, glassy phase, which can be seen under a phase-contrast microscope, indicates that the refractive index of the crystal is lower than that of the glassy matrix. Square grains substantially smaller than 250 nm can be observed in TEM microphotos magnified 20,000 times. The electron diffraction pattern shows distinct rings characteristic of crystalline phases. The elemental analysis of microprobe tests (energy dispersive X-ray spectra) indicates that the crystal grains are composed of a non-silicate substance with a high calcium content. Very small, blurred grains (about 50 A) can also be observed in TEM microphotos magnified 200,000 times. Therefore, the opalescence of the glass is the result of the presence of large amounts of CaF_2 crystals with different sizes, especially crystals smaller than 1000 A.

The color of the glass samples is identical to that of blue vitriol, produced by the coloring effect of the cupric ion. Very small brown beads were found on the original surface of the excavated glass in the pots. The main phase of the beads, shown by XRD analysis, is copper glance Cu_2S, which was formed by the reaction of the glass melt with sulfur dioxide, coal dust, or other reducing agents at high temperatures.

The weathered layer on the surface of the samples was analyzed by XRD analysis. The main crystalline phase is wairakite, $CaO.Al_2O_3.4SiO_2.2H_2O$, a product of dealkalization and hydration of the glass during its long period of burial in the earth.

Samples of Pot Refractories

Two distinct layers can be seen in the cross section of the excavated pot sherds. The inner layer is darker and coated with glass. Traces of the reaction zone with glass batch components can be seen on the undersurface of the pot cover.

The results of XRD analysis of the coarse grains and of the complete samples of the pot refractories are basically the same. Each is composed of a glassy phase, mullite, quartz, and cristobalite, together with a small amount of hematite and rutile. It appears that the clay and temper of the glass pot were prepared from the same raw materials. In addition to the crystalline phases mentioned above, cordierite (which would have enhanced thermal shock resistance) was found by XRD analysis in the pot cover sample.

The refractory level of the pot refractories sample is 1610-1630°C. The porosity of the inner layer is about 22%, while that of the outer layer is about 40%. It is believed that the variance in porosity between the inner and outer layers resulted from the different grain compositions of the temper.

XRD analysis also revealed that the main crystalline phase of the intermediate zone between the glass and the pot refractory is leucite $K_2O.Al_2O_3.4SiO_2$. This zone also contains zeolites together with small amounts of iron, calcium fluoride, and hematite. The zeolite minerals in the intermediate zone were also observed under a polarizing microscope, and feldspar was found as the reaction product of the glass melt and mullite in the pot refractories.

Our chemical analysis showed that the opal glass sample is of the potash-lime type, rich in alumina and containing fluorine. This is similar to the Boshan glasses described in the ancient literature[2] as well as in an investigation report of the 1930s.[3] The high potash in ancient Boshan glasses was introduced by saltpeter, the alumina and fluorine by feldspathic minerals and fluorspar, respectively. {Editor's Note: If one assumes that ~65-70% of the fluorine introduced into the batch was retained in the resulting glass (and the rest lost by volatilization), then the calcium introduced along with the fluorine (assuming it was in the form of fluorspar) would contribute ~9.3-9.9% CaO to the glass composition. Since the analysis shows 9.81% CaO, that means that all — or virtually all — the lime found in the Boshan specimen can be accounted for by the fluorspar. Consequently, none of the other batch ingredients could have contained more than contamination levels of calcium. See also the analysis of sample no. 4020 in Chapter 4.}

Wood ash was used as an alkali in some European glass factories in the Middle Ages. The glasses they produced had extremely high lime contents and a certain amount of P_2O_5.[4] The ancient Boshan glasses, in which feldspathic minerals were the main raw materials and fluorspar was used as both a flux and opacifier, were entirely different from glasses made in the West during that period.

Table 1

	Blue opal glass adhering to pot	Refractory of the melting pot	Refractory of the pot cover
SiO_2	58.48	61.55	55.36
K_2O	16.07	1.40	2.03
Na_2O	4.42	0.67	1.25
CaO	9.81	0.59	1.64
MgO	0.26	0.74	1.20
Al_2O_3	6.58	29.36	32.73
Fe_2O_3	0.30	4.45	3.73
TiO_2	0.23	1.13	1.66
CuO	0.81	-	-
F	4.99	-	-

Judging from the size and shape of the excavated pot fragment, the height of the glass pot was about 40 cm, and its inner diameter near the rim about 30-32 cm. The wall, about five cm thick, was composed of two layers: a dense inner layer, which was more durable against the corrosion of glass melts, and a porous outer layer having better resistance to thermal shock.

A pile of coke was found in the excavation site. Ash enclosing pieces of unburned coke filled the furnace pits. The fact that the melting furnaces of the Boshan glassworks were fueled by coke is also noted in the ancient Chinese literature.[5] In Europe, coal was not used to fuel glass furnaces until the beginning of the 17th century.[6]

Hair clasps were made from glass rods or tubes by a secondary process; they were reheated and then formed. Button-like ornaments were made by reheating and molding. These excavated glass artifacts show that forming techniques such as tube-drawing, molding, and various types of off-hand processing were used in the ancient Boshan glassworks.

About 20 furnace pits arranged in a line were found during the excavation of the foundation trenches of buildings. Although only part of the site was excavated due to various limitations, the findings show that there was an ancient glassworks there with highly-developed skills and a considerable scale of production. The origin of glass manufacture in the Boshan area must date back to much earlier times.

Notes

1. Yang Boda, private communication, March 1983.
2. Sun Tingquan, *Annals of Yan-Shan*, 1665.
3. Xie Hui, *Quarterly Journal of The Jiaotung University*, v. 13, 1934, p. 1.
4. W. Geilmann, *Glastechnische Berichte*, v. 28, 1955, p. 146, and R.H. Brill, "Scientific Studies of Stained Glass," *Journal of Glass Studies*, v. 12, 1970, pp. 47-68.
5. See note 2.
6. G. W. Morey, *Properties of Glass*, New York: Reinhold, 1954, p. 13.

CHAPTER 8

ANALYSES OF SOME QING PERIOD CHINESE GLASSES: AN INTERIM REPORT

PAMELA ENGLAND, JAMES C.Y. WATT, LAMBERTUS VAN ZELST

Museum of Fine Arts,
Boston, Massachusetts

The Research Laboratory of the Museum of Fine Arts, Boston, has investigated a number of glass objects from the Museum's Asiatic Department. All were Qing-Period pieces and included plates, boxes, and vases. Using rapid techniques, we attempted to understand the nature of the glass and to determine the opacifiers and coloring agents used. While the scope of the survey was limited, a representative of each glass color was investigated.

Using energy dispersive x-ray fluorescence spectrometry, a non-destructive technique, the surfaces of the glasses were analyzed. Generally speaking, this analytical method is not applicable to the detection of elements below calcium in the Periodic Table. However, by surrounding the object and the detector with an atmosphere of helium, it was possible to detect the presence of elements above, and including, silicon.

A second analytical tool employed was laser-excited emission spectroscopy. Because the laser causes loss of sample material in the form of a crater some 25 microns in diameter, this method was restricted to the bases of half of the pieces included in our survey. This gave additional information on the low atomic number elements. The combination of these two techniques, however, gives little more than a qualitative analysis. With reference to The Corning Museum of Glass reference standards A, B, C, and D,[1] semi-quantitative information may be inferred for some elements.

Despite the limitation of the techniques employed, a number of interesting points have arisen from this preliminary investigation. Significant amounts of lead were present in all of the opaque pieces. Potassium and calcium were also detected, although the presence of sodium can be neither confirmed nor denied at this stage.

The laser emission analysis revealed that boron is present at more than trace levels, strongly suggesting the use of a borax flux. It has been suggested that the Jesuits introduced borax into China as it is reportedly mentioned in *L'Arte Vetraria* of Antonio Neri, 1612.[2]

Of special interest was the detection of significant levels of arsenic in most of the glass objects analyzed, although it was not detected in two light-green opaque vases nor in two strongly colored (amber and amethyst) transparent vessels (Figs. 1 and 2). It seems most likely that the primary function of the arsenic was to decolorize the glass, although its functions as a fining and opacifying agent may have also been exploited. There is certainly sufficient iron in the raw materials of the glass to give it an undesirable green tinge. In other cultures, this is known to have been removed by the addition of antimony or manganese which oxidizes the iron from the ferrous to the ferric state.

Arsenic was not detected in strongly colored vessels such as a transparent amber bowl whose color is due to iron, most probably melted in the presence of sulfur and under reducing conditions.[3] The amethyst transparent vase is colored by manganese which masks any effect due to iron and is, itself, a decolorizer. The absence of arsenic in the compositions of two light green opaque vases (Figs. 3 and 4) is likewise explainable when it is realized that the color in fact derives from iron with copper.

Many of the raw ingredients of the Qing glass inferred from the above analyses have their equivalents in the records of the 18th-century Palace glass workshop recipes in which borax, lead carbonate, and arsenic oxide are listed.[4] Calcium fluoride (the mineral fluorspar) is also cited as an ingredient (2% by weight). We were unable to confirm its presence in this study as it was not possible to remove a sample for its determination by x-ray diffraction. The use of calcium fluoride opacifier has been established in 14th-century Chinese glass.[5]

From the first half of the eighteenth century Venetian glassmakers employed a lead oxyarsenate opacifier: $3Pb_3(AsO_4)_2.PbO$.[6] However, as we did not detect arsenic in the light green opaque glasses, the use of this formulation in China might be ruled out. Mr. Rooksby in his 1961 article notes the opacifying properties of lead oxyphosphate which is isostructural with lead oxyarsenate.

In Middle Eastern and Roman opaque glasses, a yellow color was achieved by the formation of a lead antimonate. Later, a yellow lead stannate was used.[7] In the absence of any other obvious colorant in the six examples of opaque yellow Qing glass tested (one of them illustrated in Fig. 5), we tentatively suggest that it may derive from an analogous lead arsenate. In some 10th-century A.D. glass found in England, x-ray diffraction analysis has demonstrated that some opaque lemon-yellow glass contains silver arsenic sulfide.[8] An English glassmaker of the late eighteenth century has left notes on his experiments in achieving a yellow glass with the use of arsenic.[9]

Other coloring agents have been determined. The deep blue is caused by a small addition of a cobalt (Fig. 6), and the blue-greens contain copper (Fig. 7). Copper occurs with iron in the light greens, as mentioned above. Copper is also present in the red glass (Fig. 8) with one exception, in which colloidal gold was found to be the colorant. A lime-green transparent piece, in the form of a dragon, contains uranium (Fig. 10). The use of uranium in Western glass dates from around 1840.

There are a few pieces of opaque glass described as "imitation tortoise-shell" in which bands and swirls of orange and red intermingle (Fig. 11). As far as our qualitative analyses go, there was little difference between the compositions of the red and yellow glass, each indicating that copper, in the presence of lead and arsenic, was the colorant. It seems possible that the two colors were separately made, possibly under different oxidizing conditions, and then melted together using a millefiori or marbling technique.

The pieces were formed by casting and by mold-blowing techniques. Many edges and surface details were then cut deeper or sharpened. Many fine details, such as the reign marks, were cut with an engraver's wheel. Although this survey presents only a preliminary investigation into the compositions of Qing glass, it has presented a number of intriguing questions including the history of the use of borax and arsenic oxide in glass formulation, and also the nature of the coloring agent of the opaque yellow glass.

Addendum, November 1985

Since completing this project, we have examined two more examples of Qing glass: BMFA 11.9706, a glass beaker, and BMFA 26.730, a glass vase. Both are translucent green, overlaid on white opaque glass. The white opaque contains arsenic as noted above. The translucent green is colored by copper, zinc and a minor amount of lead. This indicates that brass shavings were added to the glass to achieve the translucent green.

Figure 1. *Transparent amber-colored glass bowl with molded ring-handles and lotus leaf on exterior base. D. 7 cm. (26.760)*

Figure 2. *Transparent amethyst glass vase in the form of a pine tree trunk. H. 11.6 cm. (11.9666)*

Notes

1. Robert H. Brill, "A Chemical-Analytical Round-Robin on Four Synthetic Ancient Glasses," *The International Congress on Glass, Versailles, Artistic and Historical Communications,* Paris: l'Institute du verre, 1972, pp. 93-110.
2. Yi Jialiang (Personal communication.)
3. J. W. H. Schreurs and Robert H. Brill, "Iron and Sulfur Related Colors in Ancient Glasses," *Archaeometry,* v. 26, 1984, pp. 199-210.
4. Yang Boda, "A Brief Account of Glass in the Qing Dynasty," *Palace Museum Journal,* no. 4, 1983, pp. 3-17. (In Chinese.) See also this volume, Chapter 13.
5. Yi Jialiang, "The Technology of Ancient Chinese Glasses in Boshan around the 14th Century," this volume, Chapter 7.
6. H. P. Rooksby, "Opacifiers in Opal Glasses," *General Electric Company Journal,* v. 29, no. 1, 1962, pp. 20-26.
7. Robert H. Brill, "The Scientific Investigation of Ancient Glasses," *Proceedings of the 8th International Congress on Glass,* London, 1968, pp. 47-68.
8. S. Wyles and L. Biek, "The X-ray Diffraction Patterns of Some Indeterminate Archaeological Specimens," (referenced in L. Biek and J. Bayley, "Glass and Other Vitreous Materials," *World Archeology,* v. 2, 1979, p. 3).
9. Original notebooks in the Frank Wood Joint Library of Glass Technology at the University of Sheffield.

Figure 3. *Opaque light green "flower basket" type of glass vase. Wheel-cut mark of* Quinlung nien chih *in regular script within square cartouche. H. 10.3 cm. (11.9762)*

Figure 4. *Opaque light green mallet-shaped glass vase. Wheel-cut mask of* Quinlung nien chih *in regular script within square cartouche. H. 20.35 cm. (11.9701)*

Figure 5. *Opaque yellow glass vase with carved decoration on the sides; ring-handles held by elephant masks (after a bronze original). H. 16 cm. (26.602)*

Figure 6. *Transparent deep blue glass vase with cut and gilded decoration of flowering branches within double-line panels. Wheel-cut mark of* Ku-yueh-hsien chih *in regular script within a square cartouche. H. 17.1 cm. (11.9713)*

Figure 7. *Opaque blue-green glass dish. Wheel-cut mark of* Quinlung nien chih *in regular script within a square cartouche. D. 18.2 cm. (26.756)*

Figure 8. *Cameo-cut square glass brush holder. Butterflies and plants in red over opaque white ground. Wheel-cut mark of* Quinlung nien chih. *H. 12.0 cm. (11.9721)*

Figure 9. *Cameo-cut small glass rouge box with dragon-fly and orchid in red over opaque white on the lid, the base in the form of a leaf in bluish-green. D. 5.3 cm. (11.9778)*

Figure 10. *Transparent lime-green glass "dragon." H. 5.55 cm. (11.9782)*

Figure 11. *"Tortoise shell" glass small box (showing lid and interior of base). Mark of* Quinlung nien chih *in seal script within square cartouche on base. D. 7.0 cm. (11.9755)*

CHAPTER 9

PHYSICAL PROPERTIES OF EARLY CHINESE GLASSES

ROBERT H. BRILL AND JOHN F. WOSINSKI

The Corning Museum of Glass, Corning, N.Y., and
Corning Glass Works,
Corning, N.Y.

A KNOWLEDGE of the physical properties of historical glasses can sometimes be useful. For example, if an object cannot be sampled for chemical analysis, or if only very small samples are available, determinations of the index of refraction or density of the glass can be used to learn whether or not the glass contains an appreciable percentage of lead. By making certain assumptions, it is even possible to estimate what that lead content might be. Along other lines, the viscosity-temperature curve of a glass can be used to estimate what temperatures would have been required to carry out various glassmaking operations. In turn, this establishes the time-temperature conditions which must have been available to early glassmakers.

For several years The Corning Museum of Glass has been compiling an as yet unpublished library of physical properties of the chemical compositions most frequently found in historical glasses. In this paper, physical property values are presented for several chemical compositions which characterize Chinese glasses made during different periods.

Some of the values were determined on small samples removed from fragments of broken Chinese objects. The samples were later used for chemical analyses. Other values were determined for glasses melted in the laboratory; these were based on analyses of early Chinese glasses. Those glasses were melted by the late August A. Erickson and Andrew Kacyon, both of Corning Glass Works. The Chinese objects tested are in the collections of the Royal Ontario Museum and The Corning Museum of Glass. Detailed descriptions of the samples, quantitative chemical analyses, and illustrations of most of the objects are included in another paper in this Symposium.[1] The compositions of the synthetic glasses are given in Table 1 while those of the compositions of the analyzed glasses are stated in abbreviated forms throughout this paper.

The compositions studied, with some variants, are:

1. Lead:barium:silica
2. Lead:silica
3. Potash:lime:silica
4. Soda:lime:silica
5. Potash:lead:silica

Density

The densities (actually specific gravities) were determined for objects by weighing them in air and water. This can be regarded as a nondestructive, nonsampling test, providing one is careful not to alter any surface decoration or weathering products which might be on the objects. The determinations were carried out by Joseph Giardina (and his colleagues) of Corning Glass Works, and the results are tabulated in Tables 2a and 2b. The values have been reported to three significant figures, which, allowing for the fact that some of the glasses contained bubbles or seed, is probably the limit of their reliability. At that level of reliability, and considering the uses to be made of the data, the specific gravity and density were assumed to be numerically equal.

It can be seen that the heavily-leaded glasses are easily distinguished from the non-leaded glasses on the basis of their higher densities. That in itself is useful, but the data could be made far more useful if the density could be used to estimate the percentage of lead oxide present. Density is an approximately additive property, but the estimation of lead content from density is not always straightforward, because the density is also affected by the relative amounts of all the other oxides present. If these are primarily light elements, all of which make about the same contribution to density, there is little difficulty (except perhaps for lime, which makes a somewhat greater contribution). However, barium oxide, if present, contributes importantly to the density, its effect being about 75% of that of lead oxide on a weight percentage basis. In view of this, if nothing at all is known about the composition of an early Chinese glass, the situation is ambiguous. The best one can do is say that a glass with a given density contains either a certain percentage of lead oxide or a certain range of percentages of lead and barium oxides combined.

In cases where no compositional information at all is available, specific gravities of objects may prove especially useful for studying Chinese glasses of the Han Dynasty or other Asian glasses of about the same dates. If the specific gravity of a Chinese glass turns out to be low enough to eliminate the possibility that the glass contains a significant level of lead, then the glass is most likely to be a soda:lime:silica or potash:silica glass. From the limited evidence we have at hand now, it seems that the densities of potash:silica glasses are about 2.32 g/cc, that is, about 0.1 g/cc or more lower than those of typical ancient soda-limes. Thus, accurate specific gravity determinations of unweathered specimens relatively free of bubbles might be sufficient for identifying potash:silica glasses. (However, the same is not true for potash-lime glasses because their densities overlap those of soda-limes.)

Index of Refraction

Indices of refraction were determined for most of the glasses for which density measurements had been made. The determinations were made by grinding small chips of the glass to a powder and immersing them in a series of calibrated immersion oils under a microscope. White light was used instead of monochromatic sodium D light. The determinations were made by Terry Fisk of Corning Glass Works. The results are tabulated in Tables 2a and 2b.

Once again, the heavily-leaded glasses are easily distinguished from the non-leaded glasses, but the same arguments which apply to density also apply to refractive index, as far as its being used to estimate the lead oxide content of a glass. The balance of the composition has an effect on the index, and the barium oxide content affects index of refraction to about the same relative extent as it does density.

If the relative effects of PbO and BaO on density and refractive index differed sufficiently, values of the two properties could be used to decide whether the glass contains only PbO or PbO + BaO. Unfortunately, however, the index and density contributions of a given oxide component are not really independent of one another, and using them to discriminate between PbO and PbO + BaO glasses is not likely to be a very sensitive procedure. A plot of the index vs. density for 13 PbO and PbO + BaO glasses used in this study yielded a straight line fitting the equation: $n = 1.214 + 0.118d$. The closeness of the fit ($r^2 = 0.990$) indicates the interdependence of the two properties.[2] The equation can be used for estimating either property of a lead-containing Chinese glass if the other property is known, should one have occasion to do so.

Viscosity-Temperature Curve

The viscosity-temperature relationship[3] is in some ways the most fundamental physical property of a glass, because it establishes the temperature ranges in which a glass can be formed or otherwise manipulated by various glassmaking operations. In addition, the steepness of a curve determines the working range of the glass.

Complete viscosity-temperature curves were determined for two experimental glasses having high PbO and high BaO contents as well as for several other glasses. The work was conducted by Eugene Fontana of Corning Glass Works. Although the compositions of the glasses do not precisely match those of actual specimens of early Chinese glasses, they are close enough that the curves can be used as approximations for the chemical families mentioned above. The curves are shown in Figure 1. In addition, softening points, annealing points, and strain points were determined by Loren Morse of Corning Glass Works for these and several other glasses. The measurements were made by Loren Morse also of Corning Glass Works. The results are tabulated in Table 3.

Table 4 contains estimates of the viscosities required for certain basic glassmaking operations, some of which were used by Chinese glassmakers in ancient times. From the viscosity curves in Figure 1, values have been placed in Table 4 which correspond approximately to the temperatures at which early Chinese glasses would have had the viscosities required for carrying out the operations listed. For example, in order for a heavily-leaded barium-containing glass with a composition resembling those of the first group of samples listed in Table 2 to be wound around a wire (or other core) to form a bead, the glass would have to be at about 950° C. (Allowing for the fact that the glass gather would cool during the process, the starting temperature required might be perhaps 75°C hotter.)

Other Properties

Some other properties were determined for smaller groups of glasses where sufficient samples were available. These values are reported in Table 3. They include expansion coefficients, Knoop hardness values, and liquidus temperatures. The values were determined, respectively, by Patrick Gehl, Michael Wasilewski, and Dirk Sears, all of Corning Glass Works. Unfortunately, a series of experiments measuring the durabilities of Chinese glasses relative to other historical glasses has not yet been completed. When they have been, the results will be published in cooperation with Carol Jantzen of the Savannah River Laboratory of E. I. du Pont de Nemours & Co.

Notes

1. Robert H. Brill, Stephen S. C. Tong, and Doris Dohrenwend, this volume, Chapter 4.
2. Consider, for example, the Lorentz-Lorenz equation. For a recent discussion in Chinese see: Li Deyu, *Journal of the Chinese Silicate Society*, v. 12, no. 1, 1984, p. 10.
3. Robert H. Brill, "A Note on the Scientist's Definition of Glass," *Journal of Glass Studies*, v. 4, 1962, pp. 127-138.

Table 1

Synthetic Glasses Duplicating Early Chinese Compositions

	DB	*DC*	*DD*	*TWB*	C	*DF*	*DA*	*DE*	*JXV*	*Z*	*L*	*JXS*	*JXT*
SiO_2	63.5	42.0	19.7	38	34.4	30.5	72.2	66	65	55	56	73.5	75.5
Na_2O	3	2	1	3	1	1	20.5	14.5				22	
CaO	3	2	1		5	1	4.5	1	1		8	2	2
K_2O	2	1	0.5		3	0.5	0.5	1	20	12	11		20
MgO	1	0.5	0.3		3		0.5	0.5				1	1
Al_2O_3	2	2	2		0.8	1.5	1.5	1.5	1			1	1
Fe_2O_3	0.4	0.4	0.4		0.3	0.5	0.2	0.2					
CuO					1.20	1							
CoO	0.1	0.1	0.1		0.2		0.1	0.1					
BaO	10	20	30	14	12.0								
PbO	15	30	45	45	37.5	64		15.2	13	30	25		
Other					*					+		0.5	0.5
Prototype	c	c	c	c	c	1587, 1588	1562, 4104	1561, 1596	1560	1563, 1570	1581		

c Composite bracketing several early $PbO:BaO:SiO_2$ glasses.

* Analytical reference glass C containing 17 trace elements.

+ 3.0% Cr_2O_3

Table 2a

Densities and Refractive Indices of Chinese Glasses

Abbreviated composition
(rounded off except for major oxides)

Compositional Type	*Sample No.*	SiO_2	Na_2O	K_2O	*CaO*	*Other*	*PbO*	*BaO*	*Density g/cc*	*Refractive Index*
($PbO:BaO:SiO_2$)	83	40.5	2.7	0.2	1.0	0.7	35.2	19.7	3.29(W)	1.676
	93	37.3	3.8	0.4	1.9	9.7	37.5	9.40	3.70	1.676
	80	35.5	2.6	0.2	0.5	1.2	42.6	17.4		1.672
	84	42.4	3.6	0.1	0.4	1.0	33.3	19.2	3.77	1.656
	92	41.4	5.9	0.2	1.4	4.0	37.4	9.71	3.52	1.644
	95	51.3	6.1	0.1	0.4	2.4	28.3	11.4	3.15	1.602
	91	52.4	10.1	0.2	1.5	5.5	19.2	11.1	2.94	1.564
($PbO:SiO_2$)	1588	21.3	0.2	0.01	1.0	1.6	75.9		5.38	
	1587	19.6	0.2	0.01	0.3	4.9	75.0		5.22	>1.800
	82	27.9	0.2	0.2	0.9	8.9	61.9		4.93	1.798
($K_2O:PbO:SiO_2$)	1563	43.4	0.1	10.7	0.3		45.6	0.6	3.59	1.636
	1570	44.2	0.2	16.6	0.3	2.7	36.0		3.27	1.600
($Na_2O:CaO:PbO:SiO_2$)	1561	57.3	13.9	1.3	6.5	3.5	17.5		2.90	
	57						~17(?)		2.70	
($K_2O:CaO:PbO:SiO_2$)	1581	68.7	0.1	15.6	7.7	2.7	5.21		2.54	
($K_2O:CaO:SiO_2$)	5805	68.1	0.7	18.6	8.8	3.8			2.45	1.504
	5844	67.4	0.7	20.2	8.4	3.3			2.43	
+ 5.5 MgO	5839	64.4	1.0	8.1	18.7	2.3			2.56	1.514
+25.7 MgO	96	58.5	0.1	0.1	12.8	2.8			2.98	
({Na_2O, K_2}:$CaO:SiO_2$)	1585	55.1	8.8	9.8	10.6	5.3	0.37		2.49	1.506
	75	59.5	4.3	16.6	7.3	2.3			2.41	
($Na_2O:CaO:SiO_2$)	5841	52.8	9.2	1.2	16.9	9.9			2.55	1.516
Not Analyzed	5801								2.52	
	5803								2.41	
	5811								2.51	

(W) indicates partly weathered

Table 2b

Densities and Refractive Indices of Synthetic Chinese Glasses
Abbreviated Composition

Compositional Type	*Glass*	SiO_2	Na_2O	K_2O	*CaO*	*Other*	*PbO*	*BaO*	*Density*	*Refractive Index*
($PbO:BaO:SiO_2$)	DD	19.7	1	0.5	1	2.8	45	30	5.02	1.793
	C	34.4	1	3	5	6.7	37.5	12		1.686
	TWB	38	3				45	14	4.15	1.670
	DC	42	2	1	2	3	30	20	3.58	1.626
	DB	63.5	3	2	3	3.5	15	10	2.80	1.534
($PbO:SiO_2$)	DF	30.5	1	0.5	1	3	64		4.70	1.783
($Na_2O:PbO:SiO_2$)	DE	66	14.5	1	1	2.3	15.2		2.73	1.530
($Na_2O:CaO:SiO_2$)	DA	72.2	20.5	0.5	4.5	2.3			2.47	1.506
	JXS	73.5	22		2	2.5			2.46	1.508
($K_2O:SiO_2$)	JXT	75.5		20	2	2.5			2.39	1.498
($K_2O:PbO:SiO_2$)	Z	55		12			30		3.03	1.572
	JXV	65		20	1	1	13		2.65	1.520
($K_2O:CaO:PbO:SiO_2$)	L	56		11	8		25		3.00	1.568
($Na_2O:CaO:SiO_2$)	Roman	71	16	1	9	3			2.50	1.519
	Islamic	67	14	2	7	10			2.54, 2.47	

Table 3

Thermal and Miscellaneous Properties

Compositional Type	*Glass*	*Coefficient of expansion 25-300° (10^{-7}/°C)*	*Softening Point*	*Annealing Point*	*Strain Point*	*Liquidus*	*Devitrification Phase*	*Knoop Hardness KHN/100*
($PbO:BaO:SiO_2$)	DB	60.4	793	575	533	1208	tridymite & cristobalite	
($PbO:BaO:SiO_2$)	DC	~80	~700					
($PbO:BaO:SiO_2$)	DD	104.2	636	502	470	946		
($PbO:BaO:SiO_2$)	TWB	86.1						
($PbO:BaO:SiO_2$)	Ref. C							
($PbO:SiO_2$)	DF	~80	574					357
($Na_2O:PbO:SiO_2$)	DE	~94	625			847	tridymite	
($Na_2O:CaO:SiO_2$)	DA	~107	650			738	cristobalite	
($K_2O:PbO:SiO_2$)	Z	87.0						
($K_2O:PbO:SiO_2$)	JXV	99.1	687	485	444			
($K_2O:CaO:PbO:SiO_2$)	L	84.7	750	562	512			415
($Na_2O:CaO:SiO_2$)	Roman	97.4	699	529	490	930	devitrite	474
($Na_2O:CaO:SiO_2$)	Islamic	102, 85	690, 735					441, 448

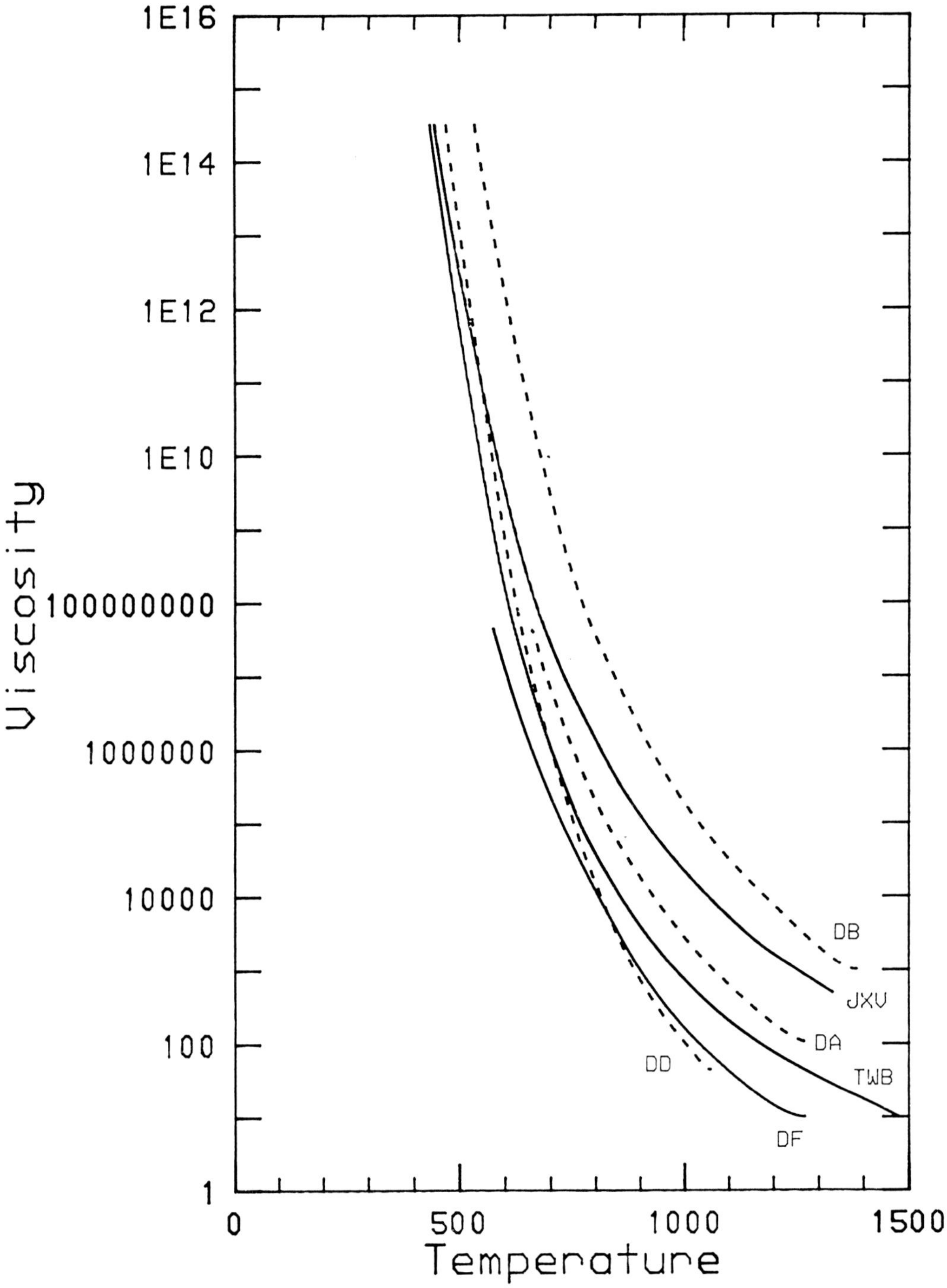

Figure 1. *Viscosity-temperature curves for some synthetic Chinese glasses. DB, DD, TWB - PbO:BaO:SiO_2; DF - PbO:SiO_2 (estimated); JXV - PbO:K_2O:SiO_2; DA - Na_2O:CaO:SiO_2.*

Table 4

Temperatures Required for Various Glassmaking Operations with Early Chinese Glasses

		Approximate temperatures			
Operation	*Log viscosity*	*$PbO:BaO:SiO_2$**	*$PbO:SiO_2$***	*$Na_2O:CaO:SiO_2$⁺*	*$K_2O:PbO:SiO_2$⁺⁺*
Melting	~3.2	~1050°	~840°	~1020°	~1030°
Gathering	~3.5	~970°	~810°	~960°	~980°
Blowing	≤4	≤950°	≤780°	≤930°	≤930°
Mold pressing	4	~950°	~780°	~930°	~930°
Fusing of grains	6	~750°	~650°	~740°	~730°
Annealing range	11-14	~520-560°	~380-430°	~470-520°	~450-530°

* Composite estimate from DB, DD and TWB.

** Estimated from DF.

$^{+}$ Estimated from DA.

$^{++}$ Estimated from JXV assuming a higher PbO content.

CHAPTER 10

A GLASS BI AND A DECORATED SWORD FROM HUNAN PROVINCE

GAO ZHIXI

The Hunan Province Museum,
Changsha, Hunan

A LARGE number of glass objects from the Chu tombs of the Warring States Period (475-221 B.C.) have been found in Hunan. Glass *bi* and sword decorations are typical of these finds. This paper presents an investigation of their excavation, use, shapes, decoration, dating, chemical compositions, and places of manufacture.

More than 120 pieces of glass *bi* have been unearthed from tombs of the Chu State in various localities of Hunan. There are more than eighty from Changsha, nineteen from Zixing, nine from Henyang, seven from Changde, five from Xiangxiang, two from Yiyang, two from Liuyang, and one from Chenzhou. Eight glass pommels from swords were found: seven in Changsha and one in Xiangxiang. Of the six glass sword sheath rings which were also uncovered, five are from Changsha and one is from Xiangxiang. In addition, there is a glass sword *bi* which was found in Yiyang.

In the *Da zhongbo* chapter of *Zhouli*, a Chinese document of the Warring States Period, there appears a passage "...to pay tribute to heaven with dark green *bi*." This indicates that the *bi* was a ceremonial object. However, archaeologists have proven that the use of *bi* was more extensive. Glass *bi* unearthed from Chu tombs in Hunan were used chiefly as decorations to be buried with the dead.

Excavated glass sword decorations include *jianshou*, *jianer*, and *jianbi*. *Jianshou* is a sword pommel, while the *jianer* is a ring-shaped decoration located near the sheath opening and used for attachment to leather belts when a sword is worn. *Jianbi* is a decoration at the end of a sheath.

Descriptions

1. Glass *bi* decoration. The predominant color of the glass *bi* from the Chu tombs is light green, followed by milk white, cream, and dark green. All of these objects are round and flat, with a hole in the middle. Statistics compiled on 57 pieces of glass *bi* from the Chu tombs in Hunan showed a smallest diameter of 6.5 cm and a largest of 14.1 cm. Most are between 7.9 and 9.4 cm in diameter. In thickness, they range from 0.2 cm to 0.45 cm with an average measurement of 0.25-0.35 cm. The lightest weighs 28 grams, the heaviest 225 grams, and the most common weights are 40-60 grams.

The decorations on glass *bi* are relatively simple and of only two kinds: grain patterns and cloud patterns. The surfaces with grain patterns are smooth and glossy while the undersides are rather rough. There are only three *bi* from the Warring States Period which have cloud patterns, and they are light green, dark green, and white (Figs. 1 and 2).

2. Glass sword decoration. The eight sword pommels are identical in shape but different in

decoration. All are light green and shaped like a round cake. The center is slightly indented. Underneath, there is a small post which would join the sword handle. There are three kinds of surface decoration. The first, found on only one of the pommels, consists of four circles in grain patterns. The second type, which is characterized by a center decoration and a fine-lined persimmon base pattern, is seen on five pommels. One of the remaining two objects has a surface decorated with a coiled dragon and grain pattern, while the other has a coiled dragon and persimmon base pattern. The diameters of the pommels are 4.2-4.7 cm, the thicknesses 0.4-0.45 cm, and the weights 18-19 grams (Fig. 3).

The six sheath rings are strip shaped and slightly curled up at both ends. Underneath is a rectangular opening for holding the belt. These rings are light green, cream, or milk white, and they bear four patterns of decoration: grain, grain with clouds, coiled dragons, and animals with grain. These rings are 6.2-10.3 cm long, 1.5-2 cm wide, and 1.1-1.4 cm thick. Their weight ranges from 19 to 39 grams (Fig. 4).

There is one *jianbi*, a cream-colored sword sheath decorated on the bottom. The surface has a grain pattern, and the object is ladder-shaped with a shuttle-shaped cross section.

Dating and Evolution

The earliest Chu tombs unearthed in Hunan Province belong to the middle or late Spring and Autumn Period (770-476 B.C.). In these tombs, a complete set of burial pottery usually consisted of the *li* (cooking vessel), *bo* (earthen bowl), *dou* (hemispherical bowl), and a jar. No glass objects were found with them. For Chu tombs of the Warring States Period, in which glass objects have been found, a complete set of burial pottery comprised a *ding* (three-or-four-legged tripod), *dui* (round or oval vessel), and *hu* (round or rectangular jar); or a *ding*, box, and *hu*. These were imitations of bronze ritual vessels. According to archaeological studies, the glass found as a funeral object with the *ding*, *dui*, and *hu* belongs to the middle or late Warring States Period (400-221 B.C.).

In general, the earlier *bi* are thin and small, while later examples are thick and large. The size of the *bi* indicated the social status of the occupant of a tomb, a larger size indicating a higher position. In some tombs with similar numbers of ritual objects, the size of the *bi* varies. This suggests that no strict provision had been made for the use of *bi* at that time.

The glass sword pommel came into being in the middle of the Warring States Period. It was popular in the later part of that period and was also in use during the Han Dynasty. The sword sheath ring has a similar history.

Chemical Compositions and Manufacturing Sites

Five fragments of glass *bi* from Hunan have been analyzed quantitatively. Another four have been analyzed by laser spectrography (Tables 1 and 2). The results show that these glasses are of the $PbO:BaO:SiO_2$ system, a type of glass unknown in the Western World.

The use of the *bi* as an important ritual object, usually made of jade or other stones, continued for more than 3,000 years from the late Neolithic age to the Han Dynasty as is known from excavated examples. Moreover, the use of *bi* was recorded in documents earlier than the Qing Dynasty, as in *Zuo zhuan* (Zuo commentary [on the Spring and Autumn Annals]). The records about *Heshi bi* and *yeguang bi* (giving light at night) dating from the Shu State are also good examples verifying that the *bi* was a traditional ritual object or a treasure. Most of the glass *bi* from the Warring States Period were unearthed from small tombs of officials and common people. Many sources provide information that glass *bi* were widely used and relatively easy to obtain in Hunan during that period. Changsha and other localities may have been centers of glassmaking.

Jade decorations on swords also have a long

history in China. A sword with jade decorations was called *yuju jian,* many of which have been unearthed in tombs of the Warring States and Han Periods. In the poem *Donghuang Taiyi* in the *Jiuge* by Qu Yuan, there is a famous remark, "Touch the jade ring of a long sword."

The grain pattern on *bi,* the coiled dragon design, and animals on swords were the prevalent decorations during the Eastern Zhou Period (770-221 B.C.). The shapes, decorations, and chemical compositions prove that these glass objects were independently made in China.

Apart from technical conditions, the emergence and development of a new product in ancient times required both an availability of raw materials and a demand for the product. These two requisites existed in Hunan. The province is rich in lead and barite mines as well as in high-quality silica resources. The enormous production of glass *bi* and sword decorations in Hunan may have been due to the fact that the province lacks jade. During the Warring States Period, a jade *bi* was considered a treasure and was used to show one's rank and standing. Swords were regarded as precious if they were decorated with jade. Lower-ranking officials and common people used glass objects in imitation of jade. The demand in society for these glass items, and the presence of the resources to produce them, must have spurred the rapid development of glass-making in Hunan in ancient times.

Notes

The author did not provide any references, however, he has apparently published the objects discussed here, along with others, in: Gao Zhixi, "Glass Wares of the Spring and Autumn and Warring States Periods," *Wenwu,* no. 12, 1985. (In Chinese.)

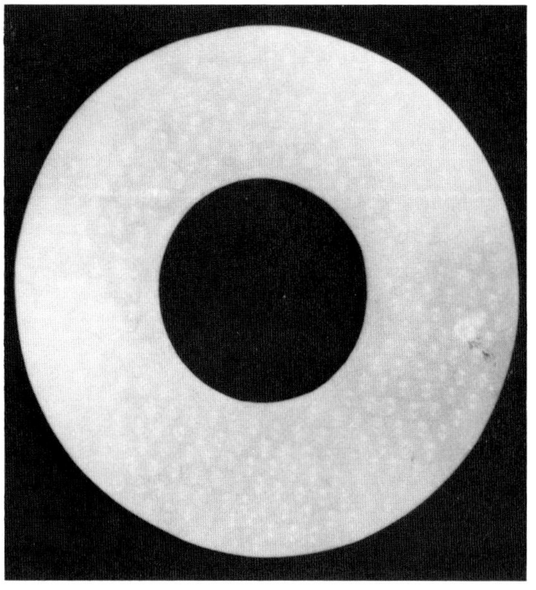

Figure 1. *Glass* bi *disk, diameter 8.2 cm, excavated in Changsha.*

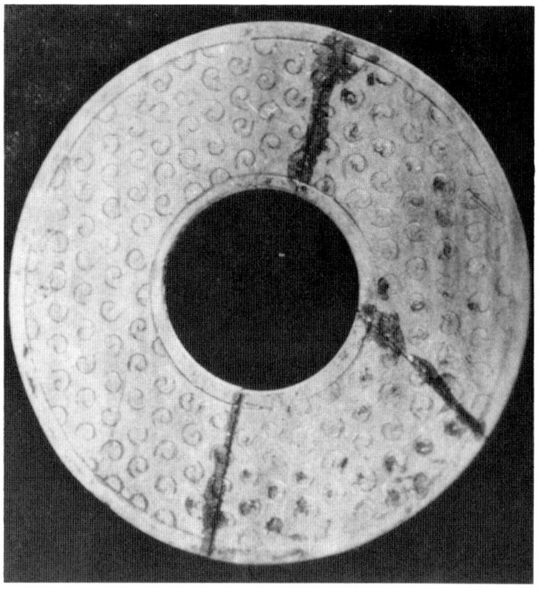

Figure 2. *Glass* bi *disk, diameter 14.1 cm, excavated in Changsha.*

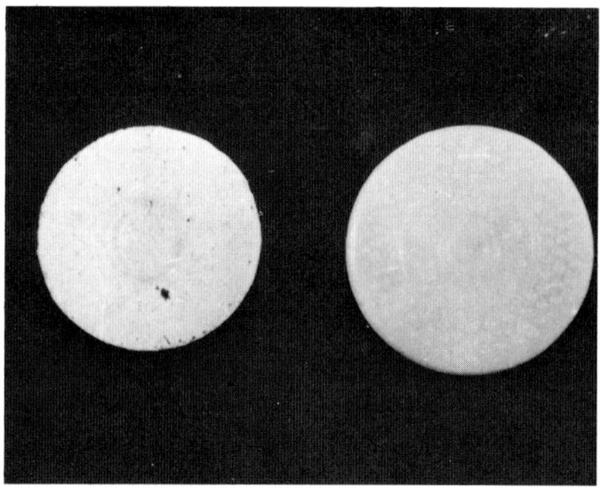

Figure 3. *Glass sword pommels* (jianshou), *diameters 4.2 cm and 4.6 cm, excavated in Changsha.* [Editor's note: *The photographs provided were not very clear, but both objects have molded decoration on the surfaces.*]

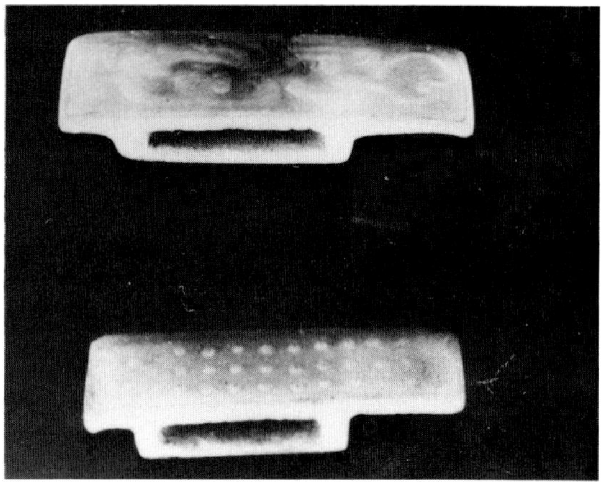

Figure 4. *Glass sword "ring"* (jianer), *lengths 6 cm and 6.2 cm, excavated in Changsha.*

Table 1
Chemical Analyses of Glasses from Hunan Province

	White *Chu tomb,* *Changsha*	bi, *Changsha*	*Lt. green* bi, *Changsha*	*Dk. green* bi, *Changsha*	*White* bi, *Hengyang*
SiO_2	39.0	34.69	37.16	38.3	36.57
PbO	28.3	37.24	39.8	41.53	44.71
BaO		10.36	13.4	10.37	10.1
Al_2O_3	0.9	} 1.16	0.62	1.67	0.64
Fe_2O_3	0.2		0.16	0.32	0.15
CaO	0.73	9.62	1.95	2.75	2.1
MgO	0.14	0.54	0.40	0.41	0.21
K_2O	0.09		0.27	0.34	0.1
Na_2O	6.4	5.2	3.32	3.07	3.72
CuO	0.02		0.03	0.73	0.02

Note: Al_2O_2 and Fe_2O_3 combined.

Table 2
Qualitative Analyses

	Yellowish bi, *Changsha*	*Opaque* bi, *Zixing*	*Lt. green* bi, *Changsha*	*Dk. green* bi, *Changsha*
Si	20	20	20	>10
Pb	10	5	5	>5
Ba	13	10	10	>5
Al	3	2	2	1.5
Fe	1	0.5	0.2	0.7
Ca	1	1	3	1.5
Mg	0.1	0.5	0.5	0.3
Na				>1
Cu	0.5			0.3
Co				0.005

CHAPTER 11

EARLY GLASSES EXCAVATED IN HUBEI PROVINCE (SUMMARY)

HOU DEJUNG

The Hubei Province Museum,
Wuhan, Hubei

JUDGING from excavated finds and historical documents, various types of small glass artifacts could have been made along the upper reaches of the Yangtze River in the 5th century B.C.

Preliminary analytical results indicate that colored glazes on the sword of Guo Jian, Prince of Yue (496-464 B.C.), probably are of the potash-lime-silica type of glass. A broken piece of colored glaze 9 mm long contained some ellipsoidal air bubbles. These colored glazes might have been pressed into shape by putting some softened glass into prefabricated models. We believe that the colored glazes were manufactured locally.

In the tomb of Zeng Houyi, more than 100 colored glass beads were unearthed. "Zeng" means "Sui," so the "Marquis of Zeng" means the "Marquis of Sui." Historical documents of the Han Dynasty refer to "the beads of the Sui Marquis." They say that the beads which the Sui Marquis fabricated by some "medicines" {or chemicals} were exquisite and as shiny as a jade pear. These colored glaze artifacts must also have been made locally. During the Spring and Autumn Period (770-476 B.C.) and the Warring States Period (475-221 B.C.), colored glazes were already being used to some extent in the upper reaches of the Yangtze River since excavated artifacts and the narrations in "The Songs of Chu" have established this.

CHAPTER 12

A NOTE ON SOME LEAD-BARIUM GLASSES FROM INDIA (SUMMARY)

H. C. BHARDWAJ

Banaras Hindu University,
Varanasi

ALTHOUGH difficult geographical barriers separate them, mutual influences occurred between Indian and Chinese cultural groups in ancient times. However, not much physical evidence has been found of contact before the Sui Dynasty, that is, ca. 600 A.D. We have made an effort to compare the chemical compositions of glass beads indigenous to India and China during the Han Period, or about the 3rd century B.C. to the 3rd century A.D. Glass beads are objects likely to have been transported through trading contacts, and it was felt that analyses of beads excavated in India might shed light on such contacts.

Our analyses show that Indian glasses of the period do not contain lead or barium, either singly or in combination with one another. On the other hand, it is well known that a combination of lead and barium is characteristic of glasses made in China during the Han Period. This is probably because barium-containing minerals occur in association with lead ores in China, particularly in Shuikoushan and in Liaoning and Hubei Provinces.

Chemical analyses of three glass beads found in Tripuri, India, show that they contain appreciable concentrations of both lead and barium. The beads are:

1. A bead of blue glass dated to 100 B.C.-200 A.D.
2. A spherical bead of transparent green glass dated to 200-400 A.D.
3. A double-anular bead of blue opaque glass dated to 200-400 A.D.

Some beads with high specific gravities (3.43) from Sulur, in the Coimbatore district, also have high lead contents and could contain barium. {Editor's Note: If this specific gravity is assumed to correspond to a density of 3.43 g/cc, the glass would probably contain approximately 41-42% PbO, without barium. If barium is present, the glass could contain approximately 30% PbO and 16% BaO; 35% PbO and 9% BaO; or any intermediate mixture. If there are air bubbles in the beads, both oxides might be somewhat higher. The refractive index of the glass should be $\sim$1.62. See Chapter 9.}

We conclude that these beads were made in China and imported into India and that they are convincing evidence of cultural contacts between India and China during this period. This topic deserves much further attention. One should also keep an eye open for glasses found in China which might have been made in India.

PART II

SUPPLEMENTARY PAPERS

CHAPTER 13

AN ACCOUNT OF QING DYNASTY GLASSMAKING

YANG BODA

Palace Museum,
Beijing

THE Qing Dynasty (1644-1911) was a period of outstanding artistic and technical accomplishment for Chinese glassmaking. A handicraft nearly 3,000 years old, Chinese glassmaking developed significantly in both production techniques and chemical composition during this period. As shown by the cultural relics and existing written works, the glassware from the reign of Qianlong (1736-1796) was made by talented and creative craftsmen. The glass objects, which were brilliant in design and color, possessed a unique artistic style different from European glass of the same time.

The Main Characteristics of Qing Glassmaking

Judging from the artifacts and ancient furnaces discovered so far, Chinese glassmaking may date back as early as the Western Zhou Dynasty (ca. 1122-770 B.C). The first glass objects were only small ornamental pieces such as beads or tubes. According to chemical analyses, the glasses were for the most part of a lead-barium composition and were blue opaque in color. Through the centuries, the process of glassmaking became more refined and better raw materials were used. A wide variety of colors for glass was developed, including blue, green, black, purple, and red. Meanwhile, glass objects were designed to fulfill practical purposes instead of being merely ornamental. By the Wei Dynasty (220-265) skillful workers apparently had learned to blow or cast colorful sacrificial vessels, daily utensils, imitation precious stones, and burial objects.

When we come down to the 17th-18th century, we can observe that all the Qing emperors had a strong liking for luxury glass items, and this greatly promoted the flowering of Chinese glassmaking. The emperors went to considerable expense to introduce Western techniques and to develop commercial exchanges between China and the outside world. Thereafter, glassworks appeared in Shaanxi, Henan, Shandong, Hubei, Hunan, Guangdong, and Guangxi, and eventually throughout the entire country.

Prior to the founding of the Qing Dynasty, glassmaking during the Ming Dynasty (1368-1644) centered around Shandong. As recorded in the *Yanshan zaji (Mt. Yanshan Records)* by the Qing writer Sun Tingquan, Yanshen was a major glassmaking area in Ming times because it was both rich in glassmaking raw materials and first-rate craftsmen. Unfortunately, in the late Ming Dynasty the town of Yanshen was struck by droughts. During the reign of Chongzhen (1628-1644), about ninety per cent of the glassworkers in the town died of starvation. Glassmaking ceased in Yanshen.

During the reign of Emperor Kangxi (1661-1722), the second ruler of the Qing Dynasty, Yanshen resumed its traditional role as a center for the glass industry. Both ornamental and practical glass objects were crafted in large quantities and sold all over China. The glassware was of such high quality that the town regained its former role as China's best glassworks.

Commercial contacts were frequent between China and the West during the late Ming Dynasty. Merchants came to China from France, Spain, Portugal, the Netherlands, and Great Britain. After a long and difficult voyage around the Cape of Good Hope and across the Indian Ocean, the foreigners landed in Imperial China's largest port cities of Guangzhou, Quanzhou, Zhangzhou, and Ningbo. These four ports soon became centers for importing and exporting Chinese and Western goods. In addition to the foreign merchants, Western missionaries also came to China. Well-versed in European technology and science, some were employed by Ming and Qing rulers.

During the reign of Kangxi, a prohibition on sea trading, issued in the early days of the Qing Dynasty, was ended. An open-door policy was put into effect, and Guangzhou became the nation's center for importing various European artworks: clocks, enamels, coral, precious stones, and glass. To meet an increasing demand by the Qing rulers for these Western artworks, local craftsmen in Guangzhou had to learn how to make them. China's earliest clocks, enamels, and glassware made in Guangzhou therefore combined both Western and Eastern features.

Even more than the Ming rulers, the Qing rulers encouraged the missionaries to teach their knowledge of Western science to the Chinese. These missionaries contributed to the development of China's astronomy, fine arts, architecture, and knowledge of geography.

In the 35th year of the reign of Kangxi (1696), an Imperial edict established a glassworks in the Qing Imperial city. Under the supervision of Western glass specialists, skilled workers from Guangzhou and Yanshen succeeded in producing overlay glasses, a relatively new technique for China.

Another important achievement during this reign was the manufacture of golden star glass, a reddish-brown glass filled with twinkling, golden specks. {Editor's Note: This sounds like aventurine glass.} The golden star glass and colored overlay technique are important highlights in the history of Chinese glassmaking.

The Qing rulers, princes, and ministers were ardent collectors of glass curios. With their financial support and a steady supply of labor, glass production at the Imperial glassworks went very smoothly for some 200 years. In the eyes of the people, glass was as valuable as precious treasures. The Qing emperors often gave glass objects to their favorite officials or sent such objects as presents to foreign envoys and rulers.

During the reign of Kangxi, some colored glasses were named after precious stones, and round, flat pieces of black glass were attached to royal gowns. During the reign of Yongzheng, formal recognition was given to the use of certain colored glasses in place of precious stones. For instance, the sapphire hat-decorations for first-ranking officials were replaced by transparent blue glass; the lapis lazuli decorations for fourth-ranking officials were replaced by blue-opaque glass, the rock crystal decorations for fifth-ranking officials were replaced by colorless glass, and the mother-of-pearl shell decorations for sixth-ranking officials were replaced by opaque-white glass.

By the end of the Qing Dynasty, the upper class still avidly collected glass as art treasures. When the property of He Kun, an important official of late Qing Dynasty, was confiscated, more than 800 glass objects were discovered in one of his storehouses.

Unfortunately, today most Qing glass art treasures are gone. The Palace Museum has only a small portion of what once existed.

Along with other industries, Chinese glassmaking declined during the Opium War. During this period, the Imperial glassworks and local glassworks in Boshan (Shandong),

Guangzhou, and Beijing suffered serious setbacks from which they never recovered.

Distribution of Major Glass Production Areas

Although glassmaking might have started in ancient China as early as the Western Zhou Dynasty, and although China was rich in raw materials, little information exists regarding the original glass producing areas. The locations of glass factories created during the long period from the Western Zhou to the Southern and Northern Dynasties (ca. 1122 B.C.-581 A.D.) remain unknown today. According to surviving historical records, the craftsman He Chou of the Sui Dynasty (581-618) produced glazed objects of porcelain in Chang'an (present-day Xian in Shaanxi Province). During the Song Dynasty (960-1279), there were glass factories in Suzhou, Chang'an, and Djan-kou, while the Ming centers were in Yanshen and areas north of the Huai River. During the Qing Dynasty, the major glassmaking centers were Boshan (the former Yanshen area), Guangzhou, Beijing, and Suzhou.

The Boshan Glassworks. Glass produccd by the people of Yanshen during the Qing Dynasty is called Boshan Glass after the old name of Boshan Prefecture. Once an important glassmaking center of the Ming Dynasty famous for its "attractive and shining glassware," Yanshen was located at Mt. Taishan and was rich in natural resources. Since there was not enough arable land to go around, most of the townspeople had to earn their living through mining, smelting iron, or making porcelain or glass.

The year when Yanshen glassmaking started remains a mystery. Judging from artifacts unearthed in the remains of furnaces and also from descriptions in *Sunshi jiapu (The Saga of the Sun Family)* written in the 14th year of the reign of Qianlong, glassmaking in Yanshen may date from the Hongwu reign of the Ming Dynasty (1368-1644). We can perhaps set the date as early as the late Yuan Dynasty (1271-1368) and leave this problem for further discussion.

After the Manchus conquered the Ming rulers and unified the entire country, the social situation became stablized. Yanshen recovered from the natural disasters of the late Ming Dynasty, and glassmaking flourished once again. Household glassworks were found both in the town and its surrounding areas. During the reign of Guangxu, about seventy per cent of the town's young people were engaged in glass blowing and casting, and their products, including beautiful screens and horizontal plaques with inscriptions for them, were sold all over China. The Boshan (Yanshen) glasses were made entirely from local materials: feldspar, fluorspar, quartz, niter, brass, iron, cinnabar, and lead. The best known Boshan glass objects were *qinglian* — curtains made up of strings of blue glass beads. Other high-quality glasses were jade-like pendants, bells, Buddha eyes, gourds, fish-shaped vases, Chinese percussion instruments, and *xuanyuan* mirrors. However, as stated by Sun Tingquan, the Qing writer of *Yanshan zaji (Mt. Yanshan Records)*, a detailed work on Boshan glassmaking, most Boshan glasswares were elaborately designed but of little practical value. In its heyday, Boshan glass spread as far as Beijing, coastal areas in southeast China, Korea, and the West.

The Guangzhou Glassmaking Center. As the most important gateway of south China to the Middle East and Europe, Guangzhou played a significant role in the development of Chinese glassmaking. As early as the Tang Dynasty (618-907), Guangzhou served as a trading center for jewelry, ivory, rhinoceros horns, spices, coral, and colored glazes. [Editor's Note: The author appears to have used *colored glazes* for the enigmatic term *liuli.*] To strengthen its control over the sea lanes, the Tang court sent an envoy there to take charge of the merchant ships. Similar institutions were set up during the Song, Yuan, Ming, and Qing Dynasties.

During the Qing Dynasty, a customs house in

Guangzhou called *Yuehaiguan* was established under the direct control of the Qing Court, and the city of Guangzhou served as an entry for importing Islamic and Western glass goods. This introduced the advanced techniques of Western glassmaking to China. According to the Ming writer Gao Lian, unlike the Qing ruling class, the Ming rulers and nobility were critical of importing Western-style glassware. In his fourteen volume book, Gao gives a detailed account of the shapes, designs, and colors of the foreign imported glass, criticizing the opaque-white, pear-green, azure-blue and yellow flasks, winecups, jars, plates, and cups as "not in good artistic taste." Gao's opinion was typical of sentiments at the time.

Glassmaking in Guangzhou may date as far back as the Western Zhou Dynasty or even earlier. In *Baopuzi (The Book of the Master Baopu)* by Ge Hong (284-364), a famous Eastern Jin Taoist monk, a description can be found of the local casting techniques for "crystal" objects. Such techniques did not originate in China but were introduced from the outside world, according to *Zhu Fan Zhi (Records on the Outside World)*, a Song work on trade between Arab countries and southeast China.

Glassmaking in Guangzhou thrived during the early Qing Dynasty. During the reign of Kangxi, some experienced craftsmen were recruited to join the technical staff at the Imperial glassworks. By the sixth year of the reign of Yongzhen, snuff bottles were produced by Guangzhou craftsmen for Court use. In the twenty-first year of the reign of Qianlong, the *Yuehaiguan* Customs House sent tribute cups with fitted covers to the Emperor to celebrate the Festival of Longevity. Unfortunately, the cups were very fragile, and the Emperor, reportedly, was irritated with the poor quality of the tribute. He ordered the Guangzhou glassworks to bear all the manufacturing costs on their own, and he refused to give them any reward.

The fragile glassware, widely called "native glass" or "Guang cast glass," was described by the Qing writer Liang Tongshu. In his work *Gu tongciqi kao (A Study of Ancient Metals and Porcelains)*, Liang wrote, "The glass cups made by the Guangzhou craftsmen are extremely fragile. When filled with hot water, they immediately burst into pieces." Liang also gave some information on glassmaking areas such as Guangzhou, Suzhou, and Qingzhou (in Shandong Province). About the types of designs and colors of the Guangzhou glass, however, he mentioned nothing.

The Qing people had a "glass fever" for Western-style glassware brought in from Europe. They spoke highly of its "durable properties and crystal-clear finish" while complaining without end about the defects in Guangzhou glass.

The Imperial Glass Factory. The Imperial Glass Factory was founded in the thirty-fifth year of the reign of Kangxi. It operated under a special agency in charge of constructing Imperial buildings and producing works of art for tribute. Normally, the leading post of the office was held by the Minister of Internal Affairs, while routine matters were handled by other officials appointed by the emperor. Under this office, there were a number of workshops and factories whose main tasks were to produce articles of different kinds for tribute. As well as craftsmen, their staff included directors, foremen, warehouse officials, and accountants. Production workers at the glassworks included first-rate craftsmen brought there from the few important glassmaking centers elsewhere in the country; they received both a salary and an allowance for food and clothing. Ordinary unskilled laborers worked at the glassworks at very low wages, and these workers were supplemented during busy times by temporary workers. In the 215 years from the establishment of the Imperial glassworks in 1696 until the fall of the Qing Dynasty in 1911, the works turned out a great number of objects. Most of the antiquities

Figure 1. *Transparent glass* cheng *bottle shaped by the gem-cutting technique and bearing an inscription:* Kangxi yuzhi.

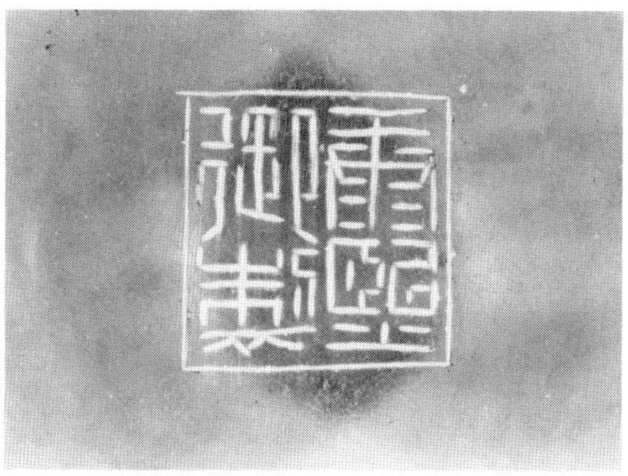

Figure 2. *Inscription of the Kangxi reign on the reverse side of the* cheng *bottle.*

Figure 5. *Yellow chrysanthemum-shaped* zhadou *(refuse vessel) made during the reign of Yongzheng.*

Figure 3. *Transparent blue vase, octagonally shaped, made during the reign of Yongzhen.*

Figure 4. *Inscription of the Yongzheng reign on the reverse side of the octagonal vase.*

Figure 6. *Inscription of the Yongzheng reign on the reverse side of the* zhadou.

found in the Palace Museum collection today were made by the Imperial craftsmen.

For a long time, the major glassmaking centers of Guangzhou and Boshan also provided an important supply of technical workers. Individuals from the technical staff were regularly chosen to serve at the Qing Court for a few months each year. During the reign of Qianlong, skilled foreign missionaries were invited to work with the Chinese. The joint efforts of the two groups yielded huge, beautiful floral lanterns made in large furnaces, as well as tributes and everyday articles made in small ordinary furnaces. According to chemical analyses, the composition of the Imperial glass was similar to that of Boshan glass, an indication that the Imperial works chiefly drew its glass from the Boshan area. At that time, glass ingots produced in Boshan were reheated by the Imperial craftsmen and blown into finished objects. During the melting and fabricating processes, the workers had to toil in high temperatures and, therefore, they would produce the annual tribute in spring, autumn, or early winter. During the summers production stopped entirely because it was too hot to work.

Generally speaking, the 200-year-old history of Imperial glassmaking reflects the major historical changes in the Qing Dynasty's power. Roughly divided into four phases, it includes the initial period from the establishement of the works to the last year of the reign of Yongzhen (1696-1735); the highest point occurred from the fifth year of the reign of Qianlong to the 24th year (1739-1758); the period of stagnation came about from late in the reign of Qianlong to the reign of Jiaqing (1759-1820); and the decline came from the reign of Daoguang to the fall of Qing rule in 1911.

As mentioned above, the main contributions of the Imperial factory to Chinese glassmaking during the reign of Kangxi were the invention of a glass-overlaying technique and "crystal." In the sixth year of the Emperor Yongzheng, the glassworks was moved to the No. 6 construction-site of the Imperial Yuanmingyuan Palaces. Early in the reign of Qianlong, when a group of Renaissance-style buildings were being constructed, the Imperial glassworkers and foreign missionaries were assigned to making huge lanterns for the buildings. This brought about significant development in glassmaking, but later in the reign of Kangxi the glassworks did nothing outstanding, the production being mainly tributes for traditional festivals.

In the reign of Jiaqing, the number of craftsmen was reduced, and the annual tribute quota of 301 glass objects could scarcely be met. After the twenty-fifth year of the reign of Jiaqing, only 160 glass objects were produced as tribute each year. In the reign of Daoguang, although the Imperial craftsmen produced 301 glass objects as tributes annually, the articles showed a marked drop in quality compared with the glass made during the two previous reigns. Craftsmen at the Imperial glassworks were brought from around Shandong, and they finished the required 160 glass objects between May and August of the Chinese lunar calendar and then returned to their hometowns. Later, two workers from Shandong came to Beijing to work in January of the lunar calendar to produce the required tributes.

The Glassmaking Center of Beijing. The traditional glass made by Beijing craftsmen is called *liao qi*, that is, objects fashioned from glass made in Shandong. The name was bestowed by the Boshan craftsmen and is now the common term for Beijing glass objects. As stated by the English writer S. W. Bushell, most of the glass used for making traditional Beijing glassware came from around Boshan. Beijing merchants also regularly bought glass ingots {or cullet}, including a kind of jade-like tile *(jing liao)*, from local people to supply Beijing workers with their needs. Glassmaking in Beijing was solely a fabrication of objects from glass ingots made elsewhere, and it did not include the initial processes of glass melting. Craftsmen reheated the ingots and cast or blew the glass into a variety of objects. These finished objects were much more expensive than the raw glass which cost only one hundred coins per *jin* (half a kilogram.). According to records, transporting glass ingots from Boshan for Beijing glass factories was a very common practice in the late Qing Dynasty and lasted until 1949.

No historical record has been found as to the exact starting date of glassmaking in Beijing. However, as Beijing was the capital for the Yuan, Ming, and Qing Dynasties, it is believed that the emergence of the handicraft there may be traced back to very early times. When taking snuff was fashionable, snuff bottles were created in many different designs and colors to meet the needs of the Qing rulers and the wealthy. During the reign of Qianlong, keen trade competition arose between the three best glassworks for snuff bottles — the Xin family, the Le family, and the Yuan family.

The Suzhou Glassmaking Center. Historical records on glassmaking in Suzhou are far from detailed. So far, it has been established only that craftsmen in the Southern Song Dynasty were able to make *sudeng,* decorative lanterns for important occasions. However, among artifacts excavated near Suzhou, an opaque-white glass *gui,* a sceptre carried by nobles at Court when they offered sacrifices, was found in the late Yuan Dynasty tomb of Zhang Shicheng's mother. Opinions vary as to whether the *gui* was actually made by workers native to Suzhou. "The glass objects (called *suzhu*) made by the people of Suzhou during the Qing Dynasty were even poorer in quality than those made by the people of Guangzhou," according to Liang Tongshu.

Chinese Glass Technology

Although Chinese glass has a centuries-old history, the records of Chinese glass technology, its compositions, and its colorants are incomplete and ambiguous. As described by Wang Chong, an Eastern Han Dynasty (25-220) philosopher, in his masterpiece *Lun heng (Discourses Weighed in the Balance),* "The Taoist monks used to make five-colored artificial jade as shiny as real jade by fusing five-colored stones .... Suihou made beads out of several 'medicines' which were shiny and appealing." When explaining the glassmaking done by the *ibin* craftsmen, the Tang writer Yan Shigu said in his book entitled *Qian Hanshu (Chronicle of the Western Han Dynasty):* "The most popular way for making glass was fusing stones and then treating the molten substance with chemicals." Nevertheless, none of the records give the names of the stones, the so-called "medicines" used, or specific details of the processes. [Editor's Note: In his book on southern China's foreign trade with East Asian countries, the Song Dynasty writer Zhao Ru pointed out that the main ingredients used in glass included lead, saltpeter* (niter), and lime* (plaster stone), and gypsum, revealing for the first time that the glass of ancient southern China included soda-lime compositions. The translations of the names of the ingredients here and in the pages which follow are, unfortunately, ambiguous. This is evident from the fact that the resulting batch recipes do not always make very good glassmaking sense. The difficulty arises from the fact that the author has translated the old Chinese terminology into modern Chinese terminology, and then had an English translation made. Each step could introduce inaccuracies, with the possibility that the second step could compound any inaccuracies introduced by the first. The editor has taken the liberty of using English terms which seem to strike the best balance between the English terms provided by the author, the editor's own interpretation of the author's equivalent Chinese terms in the author's Chinese version of the paper, and a combination of chemical and glassmaking common sense. In instances where the editor has used a term other than that supplied by the author, the term is marked with an asterisk and the author's English term follows in parentheses where it first appears in the text or is needed for clarity. Even so, room for debate still exists as to the actual ingredients described in the old documents and the subject deserves further study.}

According to *Qingzhou fuzhi (Records of Qingzhou Prefecture)*, during the reign of Jiajing in the Ming Dynasty, the Yanshen craftsmen produced glass by fusing feldspar, quartz, and fluorspar* (fluorine), and then adding litharge, white lead, and verdigris. A batch recipe used by the Yanshen craftsmen during the late Ming and early Qing Dynasties consisted of feldspar, fluorspar, quartz, saltpeter* (niter), copper* (brass), iron, cinnabar, and lead. The Yanshen batch, containing feldspar, fluorspar* (flourine), quartz, and saltpeter* (niter) as major components, was the most common and widely adopted all over the country. However, the batch for Imperial glassmaking was different. Its ingredients were feldspar (47.5%), saltpeter* (niter) (28%), borax (12.3%), *dingfen* powder (5.1%), white arsenic* (arsenic) #5.1%), and quartz {?} (2%). Compared with the Yanshen composition, the Imperial formula used during the 5th-18th years of the reign of Qianlong contained less quartz and more borax and white arsenic. It was probably influenced greatly by the German glassmaking technology then being introduced at the Qing court. After the foreign glassmaking experts, missionaries Ji Wen and Wang Zhizhong left, Chinese craftsmen abandoned this batch and resumed use of the Yanshen batch.

The most detailed description of the glassmaking process of the Yanshen people is given in the chapter *Liuli* in the *Yanshan zaji (Mt. Yanshan Records)* by the Qing writer Sun Tingquan. Born into a glassmaking master's family, Sun acquired extensive knowledge about refined glassmaking techniques as a child. Although he successfully passed the Ming's highest examinations and was appointed to be a high-ranking official in the early Qing Dynasty, Sun held an indifferent attitude towards officialdom. He left his post to write the 14-volume work *Yanshan zaji* in his hometown of Yanshen. Based on first-hand information, Sun gives a systematic and informative account of Yanshen glass raw materials, properties, coloring, melting times, tools, products, and distribution. His comments also summarize the rich working experience and creativity of the people of Yanshen. In his time, skilled craftsmen lacked sophisticated tools and instruments. They tested temperature of molten glass by examining the color of the flames. At first, various materials took a long time to melt and the charcoal flames were "black" and leaping. In the second phase, the molten substance gradually became purified and the flames appeared blue. When the melting was completed, the flames turned white, a sign that the substance was completely purified.

As they acquired more glassmaking experience, the people of Yanshen devised different chemical compositions to make glass of different colors using feldspar as the main ingredient, quartz {?} as a softening agent, fluorspar* (fluorine) as a clarifying {?} agent, and metals as colorants. The following are their ten compositions for crystal, white, crimson, blue, golden-yellow, transparent-blue, ivory-white, true black, green, and gosling-yellow ("excluding niter") glass.

1. Composition for crystal: 5 parts feldspar, 1 part quartz, and 1 part fluorspar.*
2. Composition for white glass: with less feldspar and fluorspar,* but more quartz.
3. Composition for crimson glass: 3 parts feldspar, 1 part quartz, 1 part fluorspar*, and a little copper* and iron powder added.
4. Composition for blue glass: 3 parts feldspar, 1 part quartz, no iron or fluorspar*, but with a little copper* added.

Figure 7. *Blue* sajin *(silver* star*) glass pot made in the Qing Dynasty.*

Figure 10. *Opaque white bowl with blue designs of interlocking lotuses made during the reign of Qianlong.*

Figure 8. A jiaotai *vase (glassware with layers of similar or contrasting colors) made in the Qianlong reign.*

Figure 11. *Inscription of the Qianlong reign on the reverse side of the bowl with overlaid designs.*

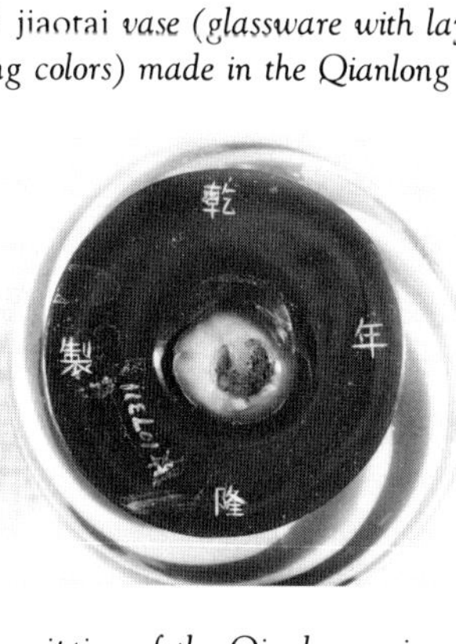

Figure 9. *Inscription of the Qianlong reign on the reverse side of the* jiaotai *vase.*

Figure 12. *Long handled opaque white incense burner overlaid with blue floral designs made during the Qing Dynasty.*

5. Composition for golden-yellow glass: Basically similar to that for white glass but with a few copper* grains added.
6. Composition for transparent-blue glass: Basically similar to that for crystal, but with a little *huawanshi* stone.
7. Composition for ivory-white glass: Similar to that for white glass, but with a large quantity of iron.
8. Composition for true black glass: Similar to that for ivory white glass, but with some iron added.
9. Composition for green glass: Similar to that for crystal glass, but with some copper* added.
10. Composition for gosling-yellow glass: Similar to that for green glass, but containing less copper* and with some sand {?} grains added.

Glass made by the Yanshen people was generally shaped by glass-blowing or glass-casting techniques. Objects which were cast included blue glass-bead curtains, glass pendants, and chess pieces, and they were finished by dropping, molding, elongating, compressing, squeezing, spinning, inlaying, or cutting. Making colored glass involved combining two strongly contrasting or similarly-colored glasses to build layered glass objects. The *chansi* glass (thread-spun glassware) was usually made by spinning five-colored glass threads evenly around the surface of glass objects. The Yanshen people also produced hollow-wares such as bulbs, fish-shaped vases, and gourds.

Glassmaking During the Reigns of Kangxi and Yongzhen

As stated by the Ming Dynasty scientist Song Yingxing in his work *Tiangong kaiwu (Exploitations {?} of Works of Nature)*, ninety percent of the glassmakers of Yanshen died during the droughts in the thirteenth and fourteenth years of the reign of Chongzhen in the Ming Dynasty. But as the social situation stabilized, the local glass craft thrived once again during the reign of Kangxi.

Qing glassmaking was based on Ming technology, and it developed rapidly, starting when the Western missionaries were employed by the Qing Court. In the thirty-fifth year of the reign of Kangxi, the Imperial glassworks was established to produce glass for court use. In the forty-fourth year of the same reign, the Emperor made an inspection tour throughout southern China and presented Song, the governor of Suzhou Prefecture, with seventeen glass objects produced by his Imperial craftsmen: an opaque-white fish-bowl, a white fish bowl, ten small yellow dishes, a blue brush-holder, a blue glass *cheng* (water container), a blue circular plate, and two *sajin* (silver-star glass) vases in white, blue, and yellow. According to experts, the complicated technique for making *sajin*-glass imitated a special precious stone from Afghanistan which seems to have silver stars twinkling against a dark blue background. {Editor's Note: This must refer to lapis lazuli. The variety occurring in Badakhshan is flecked with golden specks of pyrite or fool's gold.} The two *sajin*-glass vases of Kangxi's reign illustrate the highly developed level of the craft only nine years after the founding of the Imperial glassworks.

Another breakthrough in Qing glassmaking was the invention of a glass overlay (or casing) technique called *tao liao*. The technique included single-layer coating and also multi-layer coating, which was called *jian tao* (coating with two, three, four, or five colors of glass.) Generally, craftsmen applied colored-glass patterns over white glass. As described by the Qing scholar Zhao Zhiqian, most of the snuff bottles made in Beijing in the reign of Kangxi were opaque-white with blue or red glass patterns. But there were also some green, blue, black, or red objects clad with patterns of two, three, or five colors of glass. Sometimes, to make their objects even more elaborate, craftsmen cut floral or bird patterns into the overlays. Glasses which bore cut designs were called *kehua tao liao* (cut glass overlaid objects) while those without cut designs were called *su tao liao.*

The white objects for colored glass overlays in the middle and late parts of Kangxi's reign were of different shades of white: *chequ* shell-white, snow-white, lard-white, pearl-white, and arrowroot-starch white. White glassware coated with green, blue, or black patterns, and red ones with blue patterns, were very popular at that time. The glass overlay techniques, which might have been developed out of the *chequ*-glassmaking technique of mixing glass of two similar colors, were widely adopted towards the end of the reign of Kangxi.

Five-colored glass objects made by Kangxi's people were remarkably beautiful. The celebrated Qing poet Wang Yuyang spoke highly of them: "The white ones appear crystal-clear and ice-like, while the red ones are as red as flames." Zhao Zhiqian said, "The five-colored glasses are elegant and simple. They are rare art treasures."

However, most glass objects made during Kangxi's reign are gone today. The collection of the Palace Museum, the foremost state museum in China, renowned for its rich Qing glass, contains only one example of glass from the reign of Kangxi. It is a hexagonally-shaped crystal-glass *cheng* bottle. Following Western style, the *cheng* was blown first as a hollow bottle and then shaped by Western gem-cutting techniques. On the reverse side of the bottle, there is an inscription from the reign of Kangxi arranged in two lines of characters. It reads "*Kangxi Yuzhi.*" Judging from the unique shape and appearance of the *cheng,* it could have been done by a foreign glassworker at the Qing glassworks or under the supervision of Westerners using a Chinese jade-carving technique.

In addition to skilled workers from Yanshen and foreign glass specialists, the Imperial technical staff also included a few craftsmen from Guangzhou. Cheng Xianggui and Zhou Sun, both from Guangzhou, served at the Qing Court for more than seven years. In the forty-seventh year of the reign of Kangxi, Cheng finished a dozen cut overlay *tian qing* (azure-blue) cups for Court use and Zhou, in the fifty-fourth year, made a dozen azure cups. Both craftsmen were sent back to Guangzhou in the spring of the fifty-fourth year.

Glassmaking continued to prosper in the reign of Yongzheng. To strengthen his control of the Imperial glassworks, Aisin-Gioro Yin Zhen (1723-1735) ordered the works moved to the No. 6 work site at his newly-constructed Yuanmingyuang Palace where he held Court and attended to state affairs. According to records stored in the Construction Administrative Office under the Hall of Mental Cultivation, the Imperial glass craftsmen created charming glass objects in various shapes. There were purple glasses, ball-shaped glass clocks, carnelian utensils, red beakers, *xuanyuan* mirrors, amber-like glass cups with cut designs, strings of opaque-yellow beads, opaque-yellow *cheng* inlaid with coral designs of *chi*-dragons (without a horn) and tigers, white brush-washers, opaque-white vases with cloisonné designs of bamboo, small opaque-white *cheng,* yellow-handled cups, white *ruyi* (S-shaped ornamental objects which were once a symbol of good luck), ink slabs coated with red glass patterns, opaque-white three-footed bowls overlaid with red patterns, small jade-green bottles inlaid with enamels, red urns, and a rooster-shaped water-dropper.

The Yongzheng Imperial glassmaking workers carried on the excellent artistic tradition of the previous Kangxi reign by producing appealing objects in carnelian, jade-green, and golden amber. They also borrowed shapes from famous Ming chrysanthemum-shaped lacquerware, cups, and incense-burners with fish-shaped handles. Although the Yongzheng craftsmen added nothing new to glassmaking techniques, they did their jobs skillfully. Compared with the production processes of Kangxi's reign, their work was less refined and was sometimes criticized by the Emperor. In the tenth year of the reign of Yongzheng, the Emperor gave the craftsmen one of his favorite glass objects, an opaque-white *yu* decorated with enamel floral patterns, and ordered them to make a few copies of it. The workers tried repeatedly but failed.

A dozen glass relics exist which bear cut inscriptions from the reign of Yongzhen. They range in color from all shades of yellow to bright blue and purple. The shapes also vary. They include an octagonal vase, small jars, *shuiyu* (water-containers), *zhadou* (refuse-vessels), and round boxes. All the articles were of good quality and colorful, even though they contained a few air-bubbles and pitting, both inevitable defects in glassmaking at the time.

In the Qing Dynasty, glassware was made not only for Court use, but also as presents to foreign envoys and minority tribe chieftains. Every year, among the large quantities of presents for other countries were a large number of glass objects.

Glassmaking During the Reign of Qianlong

Qianlong's reign marked a golden age in Qing Dynasty glassmaking. Glassmaking techniques reached a peak in these days of peace and prosperity. During this period, Boshan (Yanshen) glassmaking and Beijing glassmaking both reached new heights.

In the reign of Qianlong, Yanshen became Boshan Prefecture's political, economical, and cultural center. The city's population was dense and its economy highly developed. The main street of the town, Xizhi Street, was cluttered with household glassworks which produced glass material not only for production in Yanshen but for Beijing glassmaking as well. Throughout the year, numerous merchants made trips between Boshan and the capital, Beijing, transporting glass ingots to supply Beijing workers and the Imperial glasshouses. During this unprecedented period in Yanshen glassmaking, local craftsmen regularly had to go to the Imperial glassworks. In the twenty-second year, the Boshan craftsman He Chongfu was recruited to join the technical staff of the Qing Court.

The Beijing craftsmen made their products from the glass ingots transported from Boshan. The chemical composition of the glass was similar to that of the early Qing Dynasty. The elaborately-designed snuff bottles made by the Xin, Le, and Yuan families were the best known. The jewel-like bottles made by the Xin family were translucent white or lotus-root-starch white, full of tiny air-bubbles, and clad with red or five-colored patterns. The snuff bottles made by the Le family were as white as snow and were wonderfully decorated with beautiful patterns of red and purple flowers and green trees.

Bottles made by the Le family had intricately-cut designs of fairyland pavilions, supernatural birds and beasts, and five-colored flowers. They were called *tao huadong* (peach-blossom caves). Their bodies were shell-white or translucent-white, and their blue and gold designs looked like ruby, sapphire, and gold foil. The Beijing snuff bottles were exquisite, as well as being compact and easy to carry. Coming in different designs, the smallest were only 2-3 centimeters long.

During the reign of Qianlong, Guangzhou glass was much improved from its former fragile state. Local craftsmen made their products in the traditional style, chiefly using broken pieces of imported European glassware. Their refined glassmaking techniques were admired even by some Western glass experts. Sir George Leonard Staunton, an English diplomat who visited China from 1792 to 1794, described in his work, *An Authentic Account of an Embassy from the King of Great Britain to the Emperor of China,* how ingenious the Chinese were in producing a glass substitute for a broken plate on the top of a valuable celestial globe. He said European workers

Figure 13. *Opaque white vase overlaid with traditional red designs of dragons, made during the reign of Qianlong.*

Figure 15. *Yellow glass* bo *(food container) overlaid with red floral designs, made during the reign of Qianlong.*

Figure 16. *Yellow* dou *(hemispherical fruit bowl with high stem and spreading foot) overlaid with red designs of the Chinese character* shou *(longevity).*

Figure 14. *Opaque white vase overlaid with green floral designs, made during the reign of Qianlong.*

Figure 17. *Transparent light brown bowl with ground floral designs, made during the reign of Qianlong.*

had been unable to accomplish this, and he praised Chinese craftsmen for their fine workmanship and originality. In the 21st year of Qianlong's reign, the *Yuehaiguan* Custom House in Guangzhou sent some glass mirrors, fruit-trays, and cups with fitted covers to the Qing court as tributes.

The highest point of Qing glassmaking was from the fifth year to the twenty-fourth year of the reign of Qianlong. During this period, Qing Imperial glassmaking reached unprecedented heights. In the fifth year (1740) foreigners were formally employed by the Qing Court to take part in Imperial glassmaking. The first two experts to arrive were the Jesuits Gabriel-Léonard de Broussard and Pierre d'Incarville, whose Chinese names were Ji Wen and Wang Zhizhong. They began trial-productions of *wendu lina shih*: the silver-star or *sajia* glass. The following year, after they succeeded in making this special kind of blue glass with "silver stars," they also produced a special kind of sapphire blue glass. In January of the seventh year of the reign, they accompanied the Emperor on an inspection to provide him with proposals for improving glassmaking. In the seventeenth year, since the original furnace of the Imperial works was not large enough to meet the need for melting large quantities of glass, the Emperor ordered that another even larger furnace be constructed. After the new furnace had been completed, Western and Chinese craftsmen worked together to produce three Western-style, floral-designed glasses, nine glass lanterns, eight glass urns, and two decorative glass pieces. It took them more than 100 days from November 20 of the seventeenth year till March 16 of the next year (on the Chinese lunar calendar) to complete this task. They spent more than 3,349 *liang* of silver. (There are about 32 *liang* in one kilogram.) Then they also produced three *chansi* (glass-thread spun) decorative glass objects as well as a *chansi* brush-holder.

In the twentieth year of the reign of Qianlong, the Emperor ordered the preparation of 500 snuff bottles and 3,000 pieces of glassware for his favorite officials and nobles at his summer resort in Chengde. The next year, he ordered four pairs of five-colored glass lanterns imitating the lanterns hung in the vestibule of the Shui Fa Palace. The Emperor provided more than 3,620 *liang* of silver for this work. The lanterns were designed by the talented Italian F. Giuseppe Castiglione, and they were produced under the supervision of glassmaking experts from Europe. A European Jesuit missionary mentioned the high quality, luxury products made by the Imperial glassworks in one of his letters, "Every year the works turn out an enormous quantity of vases which take a long time to shape, and are extremely costly .... Believe it or not, this glassmaking house is only one of a number of such organizations which produce luxury items for Court use."

After Ji Wen left, Imperial glassmaking began to go downhill. In the twenty-fifth year, the Emperor asked whether any new missionaries knew how to blow glass. When he learned there was no specialist at the Court, he was very disappointed. Ten years later, the *tiaotou* (a part of a lantern for holding oil) of a glass lantern which hung in the Western-style Building of the Yuanmingyuan Palaces was damaged by an official, Ji Libu. The Emperor asked the Imperial glassworks to make a replacement for the broken part. Since the old lantern had been blown by Westerners at the large glass furnace, the Chinese craftsmen could only make a poor substitute which consisted of three attached parts: an oil-holder, a pagoda-shaped decoration, and a snail-shaped decoration.

On December 7 of the same year, the Emperor ordered the craftsmen to model a few fountains after those in the Yuanmingyuan Western Building. When the man in charge of planning the project sent his drawings for the project to the Emperor, he was so worried about his technical capability that he hesitated to give the

Emperor any definite promise about the plan's feasibility. On September 10 of the next year, the work was accomplished, and it cost about 1,853 *liang* of silver. But the Emperor was displeased with the low quality of the objects and ordered the head of the Imperial storehouse, Wang De, not to give such a large sum for payment. In the end, Wang made the glassworks pay more than 156 *liang* of silver on their own. This shows that after the Western experts left, Imperial glassmaking declined in both technical capability and design.

The Qianlong glassware was very different from the simple glass objects made in the reign of Kangxi, for Qianlong glasses were very elaborate and came in many designs. Their most popular patterns showed dragons, phoenixes, birds, fish, plants, landscapes, and human figures, all engraved in a manner similar to that of jade carving.

The production of the mysterious *guyuexuan* snuff bottles was another outstanding contribution of the Qianlong craftsmen. According to Zhao Zhiqian, the bottle bore a five-colored design on its shell-white body and also displayed an inscribed poem. On the reverse side, there was an inscription of three characters indicating that it was produced by *guyuexuan* craftsmen. The *guyuexuan* bottles produced by Imperial craftsmen used to bear the character, *Qianlong yu zhi* (Made by Qianlong Imperial Craftsmen). They were even more gracefully designed. Since only a few *guyuexuan* bottles exist today and descriptions of them are contradictory, it is difficult to tell the authentic relics from fakes.

Hundreds of Qianlong glass objects have survived in the collection of the Palace Museum. However, it is difficult to tell whether they belong to the early or late Qianlong reign. Either monochrome or clad with colored glass patterns, the objects include incense burners, vases, boxes, jars, urns, pots, circular dishes, bowls, and snuff bottles. According to their colors and forming techniques, they are divided into four catagories:

1. Monochrome glassware in opaque white, shell-white, light yellow, Imperial yellow, bright yellow, bright brown, bright brownish yellow, moon white, jewel blue, azure blue, bright light blue, bright dark blue, bright crimson, bright rose, bright ruby-red, pea green, light purple, bright purple, pink, powder green, clear crystal, and brown crystal.
2. Multi-colored glass made either of *jinxing* (golden star) glass or by the *diancai* technique (decorated with colored glass designs), *jiajin* (covered with golden spots and then overlaid with glass), *jiacai* (covered with colored glass designs), and *jiaosi* (spun with glass threads of contrasting colors).
   a. Golden-star glassware was made of dark brown or reddish brown glass twinkling with golden-star lights. {Editor's Note: This sounds like aventurine glass.}
   b. Glassware made in the *diancai* technique consisted of a dark background decorated with golden dots or stars, or of a turquoise background decorated with golden-star designs. There were also red, yellow, tri-colored, or five-colored objects decorated with designs in other colors, including those with brownish-red and yellowish-green layers decorated with red and yellow dotted designs, or those with blue, white, green, and yellow layers decorated with golden star designs.
   c. Glasses made in the *jiajin* technique were decorated with golden designs, and their black or blue backgrounds were overlaid with transparent glass.
   d. Glasses made in the *jiacai* technique were first decorated with tri-colored designs in golden star, brown, green, and yellow and then overlaid with light green transparent glass.
   e. Glasses made in the *jiaosi* technique (spun with glass threads in contrasting colors)

included those with layers in purple and light purple, with light green and dark green layers, with opaque white and green layers, or with golden star brown, light yellow, and green layers.

3. Glasses made in the *taoliao* technique (glass overlay) included those with an opaque white base coated with red, blue, ruby red, or powder green designs; a transparent base coated with Buddhist blue designs; a yellow base coated with red designs; a jewel blue base coated with powder-green designs; a gray base coated with yellow designs; a pink base coated with light purple designs; and a green base coated with Buddhist blue designs. There were also examples of some opaque white glassware coated with ivory white, purple, and ruby red designs, or ivory yellow articles coated with Buddhist blue and ruby red designs, and red dotted opaque white articles coated with ruby red, transparent green, or colorless glass. After forming, some objects were either engraved, painted, etched in gold, or decorated with cloisonné enamel designs.
   a. The engraved (*diaoke*) pieces include those bearing ground or cut designs.
   b. The *miaocai* pieces are painted in bright colors on transparent or golden star glass bodies.
   c. The *nijin* (etched in gold) pieces bear golden designs on their transparent bodies.
   d. The *falangcai* pieces are opaque white and bear cloisonné enamel designs.

Like the Yongzhen glass, Qianlong glass was of excellent quality, despite having some air bubbles. Although they were only a small part of the rich Court glass collection of those days, the articles provide ample witness to the flourishing of the glassmaking craft during the reign of Qianlong.

Glassmaking During the Reigns of Jiaqing to Xuantong

From the reign of Jiaqing (1796-1820), glass production and artistic style steadily declined. In the following 107 years, craftsmen were engaged in making annual tributes for grand ceremonial occasions.

As early as the beginning of the reign of Yongzhen, (1723-1735) the Emperor established a rule that all his Imperial craftsmen should send high quality tributes to the Court to celebrate the four most important festivals of the year: the Dragon Boat Festival, the Mid-Autumn Festival, the Spring Festival, and the Longevity Festival. Neither Yongzhen nor his successor Qianlong (1736-1795) had set strict quotas on the number of annual tributes to be made. However, when Jiaqing ascended the throne, he ordered that 301 pieces of glass must be made each year. The royal craftsmen had to present eighty-one plates, cups, and small circular dishes, and sixty snuff bottles to the Emperor at the Dragon Festival and one-hundred plates, cups and circular dishes, and sixty bottles at the time of the Spring Festival. Originally, the tribute was due in two lots, in June and in October. During the eighteenth year of his reign, the Emperor ordered all tribute to be completed during May to August, and in the twenty-third year he again changed the time to February to July on the Chinese lunar calendar. All the pieces of tribute for Jiaqing's reign were made by craftsmen chosen from the Boshan area. They worked at the Imperial glassworks for a few months and then returned home after the tributes were finished.

According to the archives of the Administrative Office of Construction, Boshan craftsmen played an important role in Imperial glassmaking of the time. Two brothers from Boshan, He Zhen and He Lan, worked for the Emperor during the seventeenth to twenty-second year of Jiaqing's reign. He Lan and He Hai were later sent to the Court, and for years they worked

Figure 18. *Pink round box with incised designs of flowers and birds, made during the reign of Qianlong.*

Figure 19. *Translucent glass bowl, with painted designs, Qing Dynasty.*

Figure 20. *Small lard-white globular vase with cylindrical neck bearing cloisonné enamel floral and butterfly designs and an inscription of lines, made during the reign of Qianlong.*

Figure 21. *Lard-white vase with cloisonné designs of flowers and birds, made during the reign of Qianlong.*

there producing tribute. This system of choosing craftsmen from among local experienced people lasted for almost one-hundred years. From Jiaqing's reign on, the Imperial factory could barely make anything besides the 301 pieces of tribute each year. At times, breakdown problems occurred, and workers had to stop production for a few months or longer. They were often able to complete only half of the works required.

Broadly speaking, the glass made by the Jiaqing reign craftsmen was similar to Qianlong glass in style and technique. It is difficult to distinguish the former from the latter. With the reign of Emperor Daoguang (1821-1850), the glass started to decline in quality. Made in a slipshod way, the glass was rough and oddly-shaped. Just in color alone, Daoguang glass plummetted in quality. Inscriptions of the reign on the reverse sides of the objects were illegible. But with the reign of Xianfeng (1851-1861), the quality of Imperial glassware improved. The Emperor, reported to be angered with the quality of the tributes, ordered in his eighth year that: "All objects should be simple and all inscriptions legible."

Compared with Daoguang glass, the Tongzhi articles were much improved. But time-consuming objects, finished by gem-cutting techniques, were no longer produced. Their disappearance might be closely tied to the Xianfeng Imperial edict.

Not even the Opium War, which erupted in the twentieth year of the reign of Daoguang (1840), much affected the luxurious life- style of the emperor. The Qing ruler ordered his annual 301 tribute objects to be delivered as usual.

Meanwhile, the area around Boshan was still one of China's most important glassmaking centers. According to the Englishman Rev. Alexander Williamson, writing in 1870, glass products created by Boshan craftsmen were made from a kind of "potassium phosphate stone" mixed with saltpeter. "The objects, either of crystal-clear glass or in colorful designs, embodied the finest workmanship of the Boshan people," he said. Before the eighth year of the Tongzhi's reign, the Boshan workers were already able to produce various decorative glass objects such as gourds, windowpanes, engraved glasses, lamps, bead strings, and long glass rods. These articles were sold all over China or supplied to the Imperial or local glassworks of Beijing.

In the late nineteenth century, Boshan's economy remained unchanged. The town of Yanshen was still dotted with household glassworks. The town's skilled glassworkers, most of whom started at an early age, produced large quantities of jade-like ingots and transparent tiles for Beijing craftsmen. However, the Qing ruling class took steps to monopolize the nation's glassmaking by establishing state-run glass companies and manning them with foreign technical staffs. Strict rules were set to keep Beijing merchants from buying and transporting glass ingots from Boshan. Those who dared to break these rules would be severely punished by the government and all their goods confiscated. The restrictions brought depression to the formerly flourishing glassworks of Boshan. In the twenty-eighth year of the reign of Guangxu (1902), thousands of local craftsmen gathered to fight against the Shandong governor, Zhao Erhua, who had formed a monopolizing glass company in Boshan. They destroyed his newly founded glass company, but this uprising was suppressed by the Qing government and the three major organizers were put in prison.

Two years later, in 1904, the Qing court sent Shandong governor Hu Tinggang to establish another glass company at Liukeng, a town northeast of the Boshan. Seven German technicians employed at this company showed the local craftsmen how to produce flat plate-glass. When Qing rule was overthrown in 1911, the company was closed and the local people set up their own glass factory on the town's main street, producing flat plate-glass.

During Tongzhi's reign, the declining Qing rule underwent the so-called *Tongzhi Zhongxing (Tongzhi Restoration)* after its suppression of the Taiping Heavenly Kingdom (1851-1864), the largest peasant uprising in Chinese history. In these post-Taiping days, the political situation was relatively stable. To some extent, this promoted Beijing's local glassmaking. To meet the Emperor's unending demand for glass treasures, local craftsmen sent objects to the Qing Court every year. Glassware produced during this period was small in size but beautiful in design and excellent in quality. Some glasses modeled after sapphire and jade look very real. Today, some of these pieces can be seen in the Palace Museum.

Snuff bottles made in the reign of Tongzhi can be divided into the following groups:

1. Snuff bottles coated with colorful designs were made, including opaque white clad with moon-white designs, and dark orange ones clad with designs in brown, wax yellow, bluish green, pink, or rose.
2. The technique of painting snuff bottles on the inside was introduced in China. The inside-painted bottles made by celebrated Beijing craftsmen Zhou Leyuan and Ye Zhongsan were sent to the Qing court for tribute, and the technique spread to Boshan.
3. Bottles with molded floral designs appeared during this period. Instead of using gem cutting techniques, craftsmen made their glass by molding, a less-expensive and less labor-intensive technique.
4. Replicas of *guyuexuan* snuff bottles were made by late Qing workers. It was the habit of the late Qing craftsmen to produce replicas of relics from previous reigns. Judging from the existing relics with *guyuexuan* inscriptions found in the Palace Museum, the reproductions were lacking in quality, and they did not accurately fit the descriptions of such bottles by earlier writers. Because written records are limited, we can hardly tell the reproduction cloisonné and *guyuexuan* inscribed snuff bottles from the genuine examples.

Some Closing Thoughts

Because most of the glass products of the Qing Dynasty are gone, and only little reference data survives, I have to draw my conclusions chiefly on the basis of archives of the Imperial Constructions Administrative Office and the objects surviving in the Palace Museum. The Qing style of glassmaking, which was closely linked with its politics, economy, and culture, underwent some important changes during the two and a half centuries of the Qing rule.

Relationships Between the Imperial Glass Factory and Local Glass Factories. The establishment and development of the Imperial glassworks were possible because of highly developed local glassmaking. Regularly, the Imperial glassworks drew its technical staff and materials from glassworks all over China to meet the Imperial demand for decorative glassware under the supervision of Western glassmaking specialists. Qianlong glassmaking, which is generally considered among the highest achievements of the Qing Dynasty, surpassed local glassmaking in both quality and design during the seventeenth to the twenty-fourth years of the reign. Even during its heyday in the Qianlong reign, the Imperial glassworks still looked to local glassmaking bases for skillful workers and raw materials. Historical records are scanty and ambiguous as to the glass art in Boshan, Beijing, and Guangzhou. Therefore, I have emphasized the production of glass tributes in the Qing Dynasty. The Qianlong reign was the most important stage of Qing glassmaking.

Relationships between Chinese and Western Glassmaking Techniques. After thoroughly researching the history of Qing glassmaking, I believe that Chinese glassmaking during the Qing Dynasty was primarily based on Chinese techniques. Although for a short period the Qing court employed foreign experts to help in glass production, external help was irregular and temporary. The introduction of Western style flat plate-glassmaking only helped to improve indigenous Chinese glassmaking. In my opinion, despite its corruption, the Qing Empire fostered a prosperous native glass handicraft with its own distinct national style. To develop the quality of their tribute, all the Qing rulers were bold in introducing advanced European glass techniques. This brought about a co-existence of Chinese and Western glassmaking for a number of years.

The Qing Rulers' Influence on Chinese Glassmaking. Actually, the Qing rulers' influence is central to the relationship between Imperial and local arts and crafts. As a rule, the glassworks under the Imperial Constructions Administrative Office produced only what the emperors ordered. If their output was not made exactly to the emperor's specifications, craftsmen were rebuked or even punished. Therefore, all the Imperial glass, first and foremost, had to satisfy the rulers. Giving full play to an artisan's creative power was of secondary importance. Since imitating the Imperial glassware became a common practice of local people during the Kangxi and Qianlong reigns, local products were much influenced by Imperial style and techniques. The Boshan craftsmen would bring back home with them the tribute sketches they had made for the emperors, and they would produce copies in their local glassworks. Near the end of Qing rule, Imperial glassmaking relied on local glassmaking centers to send large quantities of tribute to the Court.

Notes

The author did not provide references with his typescript. However, there are extensive notes and references in another of his recent publications on this same subject. Many of the citations given there are related to material discussed here. Readers are encouraged to refer to the other publication which, although it treats the same subject, is not identical to this paper. Yang Boda, "A Brief Account of Qing Dynasty Glass," in Claudia Brown and Donald Rabiner, *The Robert H. Claque Collection: Chinese Glass of the Qing Dynasty 1644-1911,* Phoenix Art Museum, 1987, pp. 70-86. See also: Yang Boda, "A Brief Account of Glass of the Qing Dynasty," *Palace Museum Journal,* no. 4, 1983, pp. 3-16. (In Chinese.)

Figure 22. Guyuexuan *snuff bottle with cloisonné enamel designs of flowers and birds, Qing Dynasty.*

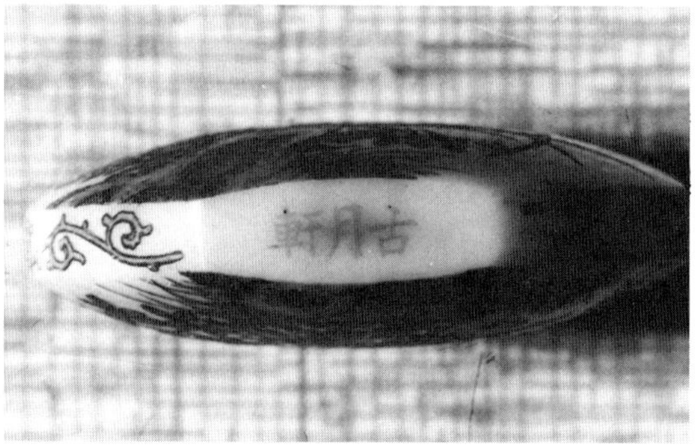

Figure 23. *Inscription of* guyuexuan *characters on the underside of the Qing snuff bottle.*

CHAPTER 14

SOME GLASSES FROM ZHOU DYNASTY TOMBS IN FUFENG COUNTY AND BAOJI, SHAANXI

WANG SHIXIONG

Shaanxi Province Museum,
Xian, Shaanxi

RESEARCHERS on the history of Chinese glass have focused much attention on the Western Zhou beads unearthed in Shaanxi and Henan Provinces. This paper reports the results of research on beads found in Western Zhou tomb groups (ca. 1000 B.C.) from Rujia village in Baoji and from Shang Song Beilu village in Fufeng, Shaanxi. The research was conducted by the Shaanxi Relics Preservation Committee.

The beads excavated from the Yu Kingdom tombs in Baoji are of four kinds: tubular, spherical, rhombic, and ellipsoidal. Figure 1 shows that some of the round beads have decorated drops while some are plain. Their colors are light green and light blue opaque. Most of the beads excavated at Beilu in Fufeng County are ellipsoidal (Fig. 2), are smaller in size, and do not have decorative drops. There are also a few tubular beads, and their colors are very similar to those of beads from Yu Kingdom tombs. Just after these beads were excavated, their colors were bright and beautiful, resembling sapphire. Compared with the beads from Yu Kingdom tombs, the Western Zhou beads from Beilu are rather simple, but they are more transparent, and their walls are thinner.

All the beads were found on the neck, breast, and arms of the deceased. They were located side-by-side with carnelians and gems, so their decorative nature is clear. Some of the beads from Yu Kingdom tombs have been consolidated with organic silicone resins after excavation, and, although the strength of these beads has been improved, their colors differ from the untreated beads. Because the beads from Beilu have not been treated, they are still quite fragile. Their color has faded to a light blue, and the outer walls have a white, weathered layer.

An important question about these beads is where were they produced? Dr. Robert H. Brill, who visited the Shaanxi Province Museum after the International Symposium on Glass in 1984, examined some of the Yu Kingdom beads. "If I had been in my New York office instead of here in China," he said, "I would have taken the tubular beads to have come from Iran or Egypt, because they have the same appearance and color as faience beads I have seen from those places. Some of the spherical beads, however, are a little different in their internal cavities, and the beads with the protrusions are unfamiliar to me."

To determine the origin of these beads, we did a lead isotope analysis on the beads from Yu Kingdom tombs. The samples were from fragments of the tubular, rhombic, and spherical beads. The results are reported in Table 1.

From lead isotope analyses of about fifty examples of ancient Chinese glass, Drs. Barnes and Brill put forward criteria for determining the origin of ancient glass by lead isotope ratios.[1] The lead isotope ratios of the beads from Yu Kingdom tombs have ratios similar to those of four Han Dynasty samples they analyzed, but they also resemble the leads in some later Chinese glasses which might have been made in the same regions. This type of lead appears to be quite common in China.[2] It should be noted, however, that while these lead isotope findings are consistent with a Chinese origin for the beads, they do not constitute absolute proof, because some leads of this isotopic type also occur in Europe. Nevertheless, they certainly do point strongly toward a Chinese origin.

The microstructure of the beads from Yu Kingdom tombs and the Beilu Zhou tomb is quite different from the faience of Egypt. The microstructure of much Egyptian faience is such that the perimeters of the quartz grains are in a glassy state while the center is crystalline. Polarized micrographs of the Yu Kingdom and the Western Zhou beads are shown in Figure 4, and that figure illustrates that the quartz particles of these beads have sharp edges. There is no sign of a melted glass phase at the edges. Glass is present as a cementing material between the different-sized quartz particles.

Among the beads from Yu Kingdom tombs, the ellipsoidal bead with decorative drops is the only type found in that area. Its decoration is stylistically similar to the nipple pattern on bronze objects excavated in the same area. This pattern is a common decoration on bronze objects of the Shang and Zhou Dynasties and the resemblance suggests that this specially decorated bead was produced locally.

The number of beads excavated and the places where they were found also imply that they were produced locally. There are more than a thousand pieces in the Yu Kingdom tombs; each of about 400 Western Zhou graves in Beilu contained anywhere from a few to several dozen beads. Since the latter group were graves of poor people, almost no bronze pieces were found in those tombs.[3] These beads were widely worn by ordinary people, and it is therefore unlikely that they were imported from distant countries. Thus we tentatively conclude that these beads were made in China. We will continue our study for conclusive results on the exact place where they were produced.

Another important question concerns how the beads were made. There are two points of view: one is that slag of smelted bronze was smelted at lower temperatures, with clay as an additive.[4] The other is that the beads were shaped by rolling the quartz granules with flux and a copper coloring agent and were then heated.[5] This question may be approached from another angle. During the Zhou Dynasty, there were two advanced handicraft technologies: bronze smelting and the manufacture of pottery and porcelain. The beads found in the tombs of the Zhou Dynasty may be viewed as products of the combination of these two technologies. Chemical analyses show that in addition to their very high silica contents, the beads also contain high percentages of copper (see Table 2), and these are not direct products of the ceramics of that time. The glazes of light blue porcelain were formed with iron {?} as a coloring agent and are also rich in lime.[6] This chemical composition is quite different from that of the beads. The high copper content of the beads leads us to believe that the techniques of bead making and bronze smelting were closely related.

There is a considerable amount of green powder scattered over the excavation of a bronze casting site of the early Shang Dynasty. An examination revealed that the powder contains quartz and a glassy phase, indicating that both weathering and smelting had occurred.[7] The discovery of this green powder shows that the process of copper smelting at that time produced by-products which were green in color and which contained a glassy phase. We can imagine that, because of accidental use of an incorrect amount of charcoal, some of the ores oc-

Figure 1a. *Yu Kingdom tubular beads.*

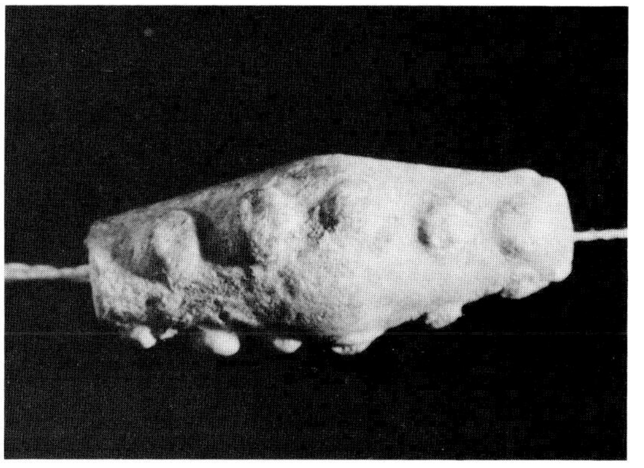

Figure 1d. *Ellipsoidal beads.*

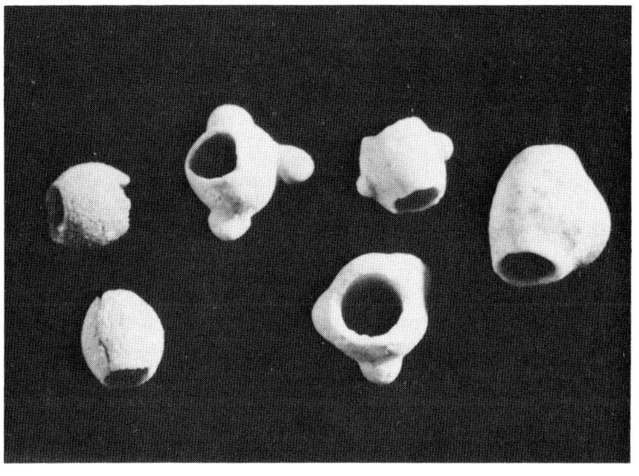

Figure 1b. *Spherical beads.*

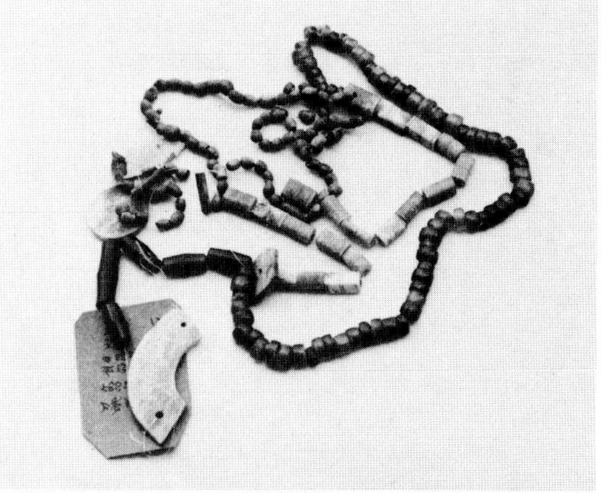

Figure 2. *Zhou Period beads from a tomb in Beilu, Fufeng County (upper left).*

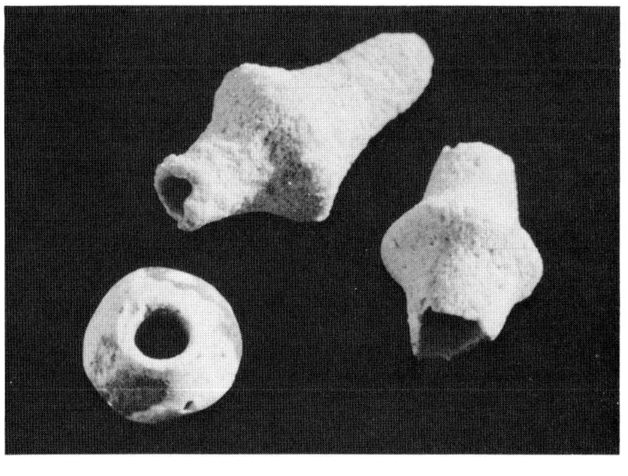

Figure 1c. *Rhomboidal beads.*

casionally were not transformed into copper and slag, but became instead a new vitreous material with a colorful appearance. This would have attracted interest and eventually could have led to intentional use for decorative purposes.

The large proportion of sharp-edged quartz particles in the beads may have improved their strength. In ancient China, it was a common practice to add sand particles to clay to strengthen pottery and improve the ease with which it could be shaped. Pottery containing sand was made in the Cishan-Peiligang culture, so the use of quartz powder as an additive seems to follow logically.

A careful examination of the cross-sections of the four types of Yu Kingdom beads (Fig. 3) showed that the walls of all but the tubular beads are thick in the middle. Thus they cannot have been made by rolling. Particularly interesting are the rhombic beads, whose shapes are extremely irregular. Some are thin and some are thick while others are long and slender. Some beads have two tubular sides and a middle portion shaped like a plate, while other beads have a spherical middle section. All of these shapes are shown in Figure 3c. The irregular shape of the rhombic beads makes it clear that they were formed at high temperatures. If they had been shaped at ordinary temperatures before melting, the workers would have made them more regular in form. The irregular shape of the beads can be explained by the inability of the workers to master the techniques of high-temperature shaping. In addition, most of the rhombic beads are not evenly formed from thin to thick. Instead, they change abruptly so that the outer profile of the cross section becomes concave. This may be due to their having been press-formed by some tool before they were cooled, and it is another indication that the beads are closely related to copper smelting. Judging from the number excavated, they must have been produced in large quantities.

Next, a question arises as to the material used for making the beads. Were they fashioned from natural or man-made substances? To answer these questions, we carried out technical examinations. The results were as follows:

1. *Specific gravity.*
   Beads from Yu Kingdom tombs: 2.24 g/cc.
   Beads from Western Zhou tombs: 2.23 g/cc.

2. *Microscopic examination and orthogonal polarized microphotography.*
   Sample: Tubular beads from Baoji Yu Kingdom tombs.
   Quartz: Broken state, sharp angles, maximum dimensions 0.02-0.15mm. 90% quartz particles.
   Glass: Uniform, light blue, index of refraction less than quartz. The glass acts as a cementing material (Fig. 4a).
   Sample: Ellipsoidal beads from Beilu Western Zhou tombs.
   Quartz: Small particles, broken, unequal size, particles are cemented by glass. There are a small number of holes. 80% quartz particles.
   Glass: Uniformly distributed, index of refraction lower than quartz (Fig. 4b).

3. *X-ray diffraction.*
   Sample: Fragment of tubular beads from Baoji Yu Kingdom tombs. Identified as quartz.
   Sample: Fragment of ellipsoidal bead from Beilu Western Zhou tombs. Identified as quartz.

4. *Quantitative chemical analysis.* (See Table 2)

These beads have a high porosity, especially those from the Beilu Zhou tombs, but the air spaces are barely visible to the unaided eye.

This is further proof that the beads are made of an artificial material, because natural minerals would not have this much porosity.

Judging from the results of the lead-isotope analysis, microscopic examinations, the bead shapes, the quantity of the beads, and the status of the wearers of these excavated Western Zhou Dynasty beads, we tentatively conclude that they were produced in China. The manufacture, and probably also the invention of these beads, was closely related to the smelting of copper. The production processes were similar to glassmaking because they included high-temperature smelting, the use of crushed quartz, shaping, and cooling until the beads were solidified. Quartz was the main constituent of these beads with a glassy phase serving as a cementing material. The way they were made suggests that these beads are linked with the origin of Chinese glass. Thus we like to regard them as a precursor of Chinese glass.

Acknowledgements. The author thanks Hu Zhisheng who took the photographs for Figures 1a-1d, and Hou Ruobin for the photograph in Figure 2.

Notes

1. I. Lynus Barnes, Robert H. Brill, and Emile C. Deal, "*Lead Isotope Studies of Early Chinese Glasses,*" *Chapter 5, this volume.*
2. *Ibid.* See also Kazuo Yamasaki and Masayo Murozumi, "Similarities Between Ancient Chinese Glasses and Glasses Excavated in Japanese Tombs," Chapter 6, this volume.
3. The Fufeng County Museum, "Excavation of a Zhou Dynasty Tomb at Beilu in Fufeng County," *Cultural Relics*, no. 7, 1984, pp. 30-41.
4. Yang Boda, "A Preliminary Investigation of Western Zhou Glasses," *Palace Museum Journal*, v. 2, 1980.
5. Zhang Fukang, Cheng Zhuhai, and Zhang Zhigang, "An Investigation of Ancient Chinese *Liuli*," *Ceramic Shul Pao*, v. 11, no. 1, March 1983, pp. 62-76.
6. Chinese Ceramic Institute, *The History of Chinese Pottery and Porcelain*, p. 79.
7. Editorial group, History of Chinese Metallurgy, *A Short History of Chinese Metallurgy*, Beijing Institute of Steel and Iron, p. 25.

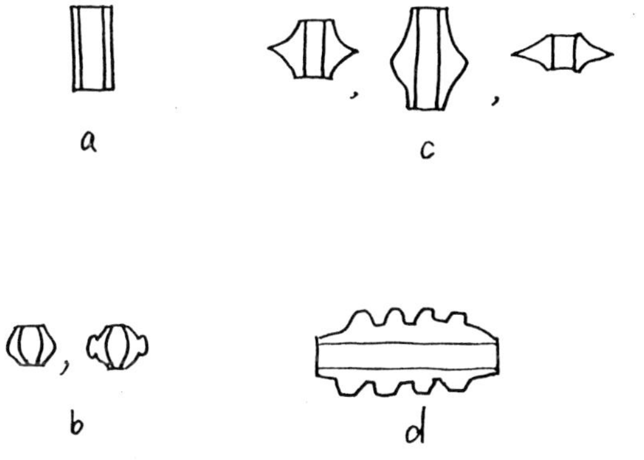

Figure 3. *Cross-sections of Yu Kingdom beads.*

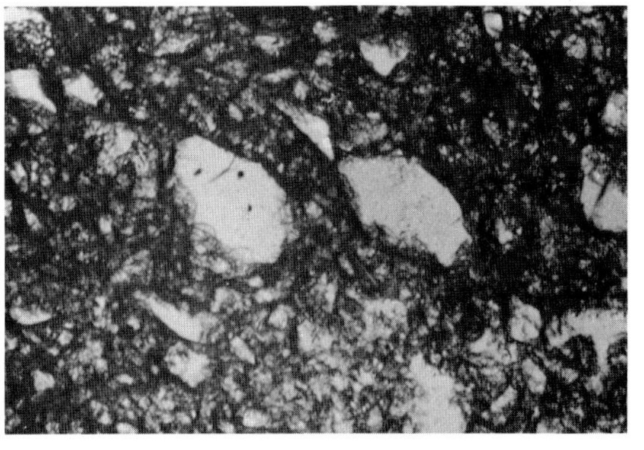

Figure 4a. *Photomicrograph of Yu Kingdom bead (~100X).*

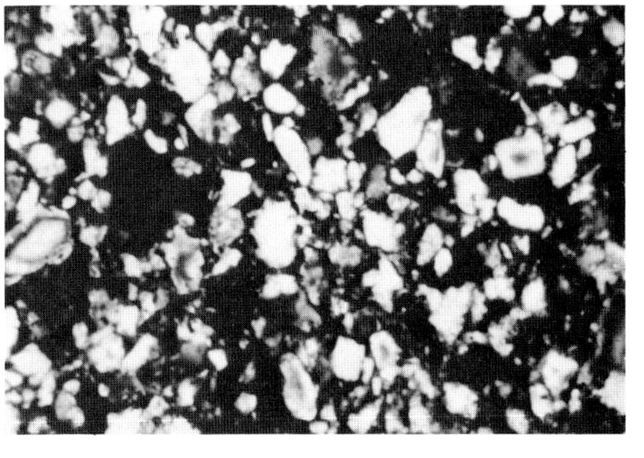

Figure 4b. *Photomicrograph of beads from tomb in Beilu (~80X).*

Table 1

Lead Isotope Ratios of Some Early Chinese Glasses

Date	*Object*	$^{207}Pb/^{206}Pb$	$^{208}Pb/^{206}Pb$	$^{204}Pb/^{206}Pb$	*Source*
Early Zhou	beads	0.8435	2.0727	0.05520	This study
Prob. Han	large bead (no. 51)	0.8468	2.1026	0.05372	Barnes *et al.*, this vol.
Prob. Han	glass fish (no. 57)	0.84730	2.09487	0.053941	*ibid.*
Han	glass cicada (no. 64)	0.84665	2.09833	0.053828	*ibid.*
Han	ear spool (no. 82)	0.84332	2.09396	0.053581	*ibid.*

Table 2

Chemical Analysis

	Beads from tomb of Beilu Zhou	*Beads from Yu Kingdom tomb*
Cu	0.6244	0.9191
Pb	0.0833	0.1283
Ba	0.0394	
Zn	0.0079	0.0068
Cr	0.0146	0.0180
V	0.0034	0.0023
Ti	0.0056	0.0158
Co	0.0006	
Mn	0.010	
Be	0.0003	

{Editor's Note: The copper:lead ratios are 7.50 and 7.16 for the above samples. Thus, if the copper colorant was a material derived from a bronze (such as oxide scale) the parent bronze alloys would have contained, respectively, 10.6% and 11.0% lead. (This is figured as a ternary alloy with an estimated tin content of 10%, since no figure is reported for tin in the analysis.) A parent bronze alloy of 79% Cu: 10% Sn: 11% Pb is a quite reasonable composition for Zhou Period bronzes. We would not regard it as a very good match for Western bronzes of *ca.* 1000 B.C., although the possibilities deserve more detailed consideration. On the face of it, the composition reported in the table with respect to the copper and lead, seems consistent with a Chinese origin for the parent alloy, and, therefore, for the beads themselves.}

CHAPTER 15

SCIENTIFIC STUDIES OF EARLY GLASSES EXCAVATED IN CHINA

ZHANG FUKANG

Shanghai Institute of Ceramics, Academia Sinica, Shanghai

CHINA has a long history of ceramic manufacture. Pottery making can be traced back to the 6th millennium B.C. or even earlier. During the Shang Dynasty (16th-11th centuries B.C.), many important technical achievements were made, including the use of high alumina clay and porcelain stone as ceramic raw materials, the emergence of kilns which could be fired as high as 1200°C, and the invention of ash glazes. High-fired and glazed ceramic wares with low porosity were successfully made at the end of the Shang Dynasty. Then the ceramic industry developed rapidly, and kilns were set up in most provinces of China. Compared with our knowledge of Chinese pottery and porcelain, relatively little is known about the origin and development of glass manufacture in China. However, this situation has been gradually improving in the past few decades, particularly in the past few years.

Since 1950, much early glass has been excavated in many provinces of China; small artifacts, such as beads, eye-beads, *bi* disks, sword terminals, cicadas, small bottles, seals, hairpins, ear pendants, belt hooks, bracelets, figurines, and glass garments for funeral purposes have been found. Most of the objects came from tombs of known dates; therefore, the remaining questions are those of places of manufacture.

Through cooperation between archaeologists and ceramic scientists, scientific research has been carried out in the Shanghai Institute of Ceramics, the Palace Museum, the Beijing Institute of Building Materials Science, the Shanghai Institute of Optics and Fine Mechanics, etc. Some of the most interesting results are summarized here.

The Beginning of Glassmaking in China. Figure 1 shows some beads excavated from a Western Zhou tomb (11th-8th century B.C.). These are the earliest synthetic beads ever discovered in China. They are about 5.8 mm in diameter and 5.5 mm high, with a bore of 2.8 mm. They are greenish and heavily weathered. In the past few years, the origin and nature of these beads have been hotly discussed. Some Chinese scholars[1,2] have concluded that the Western Zhou beads are glass, and that the beginning of glassmaking in China can thus be traced back to 11th-8th century B.C. They believe that these beads were made by melting a mixture of copper slag and clay, and they reason that glassmaking in China had a close connection with the ash glaze made by Chinese potters of the 16th-11th centuries B.C.

In order to confirm the above arguments, a study was carried out at the Shanghai Institute of Ceramics, but quite different results and conclusions were obtained.[3] Microstructure studies showed that the Western Zhou beads were composed mainly of crystalline minerals (Fig. 2).

X-ray analysis showed they contain quartz grains. The SiO_2 content varies from 90-94% according to chemical analyses, and the green color is due to copper (nos. G1, G6, Table 1). At the grain boundaries, a glass phase high in SiO_2 and alkali was found. This phase accounts for approximately 10-15% by weight. Based on these experimental results, it is clear that the Western Zhou beads are not glass, so the former suggestion that glassmaking began in China in the 11th-8th century B.C. is not correct. The Western Zhou beads were neither made of ceramic glaze nor copper slag; they were made of silica sand and a small amount of alkali. {Editor's Note: In a more recent study[4], chemical analyses of the crystalline phase and of the interstitial glass has confirmed that the beads are made of faience and revealed that the glass phase contains more potash than soda.}

Beads of the same material were also excavated from a tomb of the 8th-5th centuries B.C. in Henan Province. These are conical in shape and 12 mm high. Only the surface is greenish, while the core is white. The chemical composition is similar to the Western Zhou beads (no. G7, Table 1). Another example similar to the above is a tubular bead excavated from a Western Zhou tomb at Luoyang, Henan Province. It is 12.8 mm long and also greenish. A chemical analysis indicates that it was made of the same material as the above beads (no. G14, Table 1).

Where did these polycrystalline silica beads come from? Were they made by the Chinese or were they imported? I believe they were imported from the Middle East. As far as chemical composition is concerned, the beads are similar to Middle Eastern faience. The only difference is that most Middle Eastern faience beads are glazed, while the beads excavated in China are not; but this could be due to weathering. Faience was popular in the Middle East, but unknown in China other than for these beads.

According to historical literature, trade between China and the Middle East began during the Han Dynasty. The discovery of faience beads in China provides evidence that as early as in the 11th-8th centuries B.C. there was already trading activity, either direct or indirect, between China and the Middle East.

PbO-BaO-SiO_2 Glasses. Early glass objects excavated in China are quite varied in their chemical compositions, belonging to the PbO-BaO-SiO_2, PbO-SiO_2, Na_2O-CaO-SiO_2, K_2O-CaO-SiO_2, and mixed alkali-SiO_2 systems.[5] Most scholars agree that the beginning of glassmaking in China can be traced back to 5th-3rd centuries B.C., and that the first glass belonged to the PbO-BaO-SiO_2 system. I have analyzed a small fragment of a *bi* disk excavated from a 5th-3rd century B.C. tomb. The non-weathered core contains 41% PbO and 14% BaO. As a comparison, the weathering crust contains 56% PbO, but surprisingly little BaO, only 0.3% (no. G8 and G8W, Table 1). About 120 pieces of *bi* disks have been excavated in China. Some have been identified as PbO-BaO-SiO_2 glass.[6] According to historical records, *bi* disks were traditional ritual artifacts used between the 16th century B.C. and 3rd century A.D. Because they were decorated with traditional Chinese motifs, they were definitely made in China.

Figure 3 shows a tubular bead excavated from a 5th-3rd century B.C. tomb in Sichuan Province. It is 18 mm high, 8.5 mm in outside diameter, and slightly greenish. It is also a PbO-BaO-SiO_2 glass, but contains more Na_2O than ordinary PbO-BaO-SiO_2 glass (no. G21, Table 1). Figure 4 is a photomicrograph of that bead. The solid particles are unmelted quartz grains. The presence of so many quartz grains gives the glass poor transparency.

Figure 5 shows a large eye-bead excavated from a 5th-3rd century B.C. tomb in Henan Province. It is 30 mm high, 21 mm in outside diameter, 13 mm inside diameter. It is also a PbO-BaO-SiO_2 glass (no. G41, Table 1). The rectangular plates shown in Figures 6 and 7 are parts

Figure 1. *Left to right: Western Zhou beads (similar to G1 and G6); eye beads imported into China (see also Figure 12); fragments of small bottles, G9 and G10.*

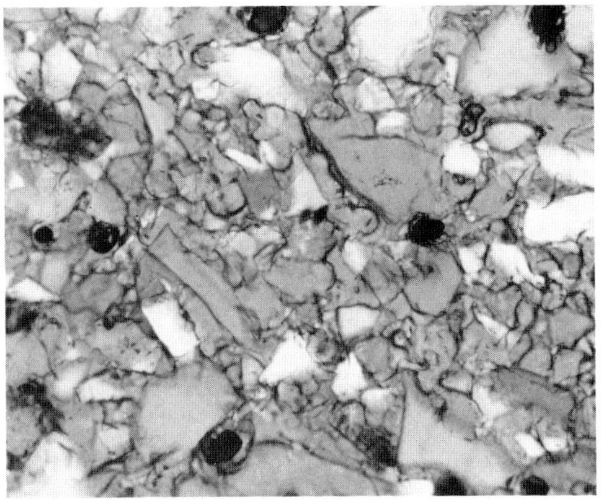

Figure 2. *Photomicrograph of G1 (~200X).*

Figure 3. *Tubular bead, G21.*

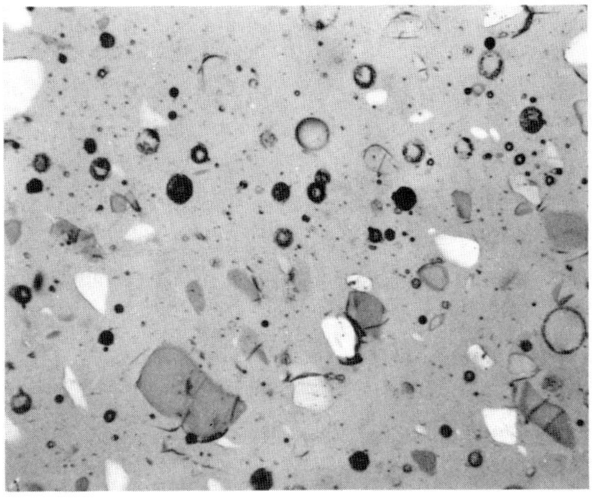

Figure 4. *Photomicrograph of G21.*

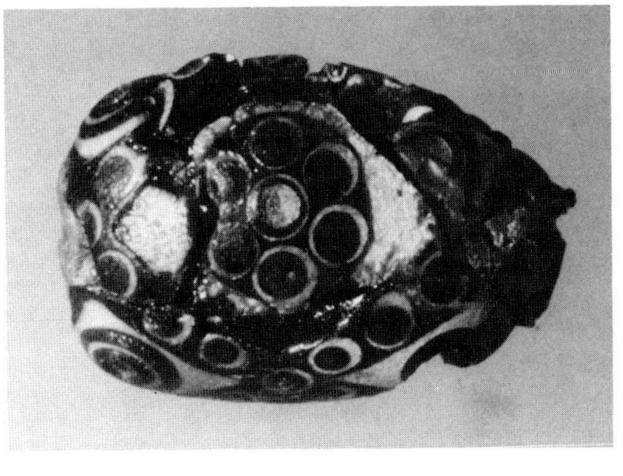

Figure 5. *Large eye bead, G41.*

Figure 6. *Pieces of glass garment, similar to G52 and G53.*

of a funeral garment excavated from a 3rd century B.C. to 1st century A.D. tomb at Yangzhou, about 200 miles northwest of Shanghai. About 600 such plates were excavated there. Some were heavily weathered. Most are rectangular, only a few are circular or trapezoidal. Figure 8 shows the different shapes. Each plate has a hole in each corner. The suit was made by joining these plates with copper wire. Chemical analyses show that the plates are made of PbO-BaO-SiO_2 glass (nos. G52 and G53, Table 1). Some of the plates, which were cast in molds, are decorated with traditional Chinese geometrical designs, as illustrated in Figures 9 and 10.

Compared with the non-weathered glass cores, the weathering crusts have higher ignition losses (approximately 10-16%), higher P_2O_5 contents (approximately 5%), relatively higher CaO contents (approximately 1.5%), a little chlorine and sulfur, and very low BaO (nos. G8, G51, G52, G53, Table 1). The high ignition loss can be attributed to the decomposition of $PbCO_3$ and loss of moisture. Phosphorus, calcium, chlorine, and sulfur were mainly derived from the decomposition products of the body. The very low BaO content, as compared with the non-weathered glass, can be attributed to the solubility of the barium compounds.

Figure 11 shows the microstructure of a glass plate. The needle-like crystals are barium disilicate, as identified by X-ray diffraction analysis.[7]

PbO-SiO_2 Glasses. The small greenish and yellowish bottles shown in Figure 1 are PbO-SiO_2 glasses. These were excavated from a 10th-13th century tomb in Anhwei Province. The PbO is approximately 67%, and the SiO_2 28% (nos. G9 and G10, Table 1). It is quite possible that this glass was made by the Chinese themselves.

Na_2O-CaO-SiO_2 Glasses. The eye-bead in Figure 1 was excavated in Henan Province in an 8th-5th century B.C. tomb. The body is a copper green glass, the eyes cobalt blue and opaque white. All are Na_2O-CaO-SiO_2 glasses. The opacifier for the white opaque glass is an antimony compound, but in China antimony was not used in the ceramic industry until the 17th century. The cobalt in this bead is of the low manganese type, leading the author to believe that this bead was not made in China. Figure 12 is a photomicrograph of the opaque white glass in this bead. In recent years, other Na_2O-CaO-SiO_2 glass objects have been excavated in Yangzhou, dating to the 8th-9th centuries, and Figures 13-19 show some of these. Yangzhou was an important trading center during the 7th-10th centuries, for at that time Islamic people came from Iran and settled in this city. In the archaeological sites, turquoise blue pottery fragments and many other relics characteristic of the Tang Dynasty were found in the same levels. Analysis and microstructure studies of this unusual pottery show that they were imported into China by Iranian merchants in the 8th-9th centuries.[8] The Na_2O-CaO-SiO_2 glass objects are also believed to have been brought in by them.

K_2O-SiO_2 and K_2O-CaO-SiO_2 Glasses. There have been many K_2O-SiO_2 glass relics excavated from Han Dynasty tombs located in Guangxi, Guangdong, Yunan, Hunan, Gansu and Jiangsu Provinces. They include beads, eye-beads, pendants, and cups. All are quite similar in chemical composition, the K_2O contents being 14-17%, SiO_2 71-81%, and Al_2O_3 2-5%. Na_2O and CaO are usually very low. Some Chinese scholars believe this type of glass was made by the Chinese. Their opinion is based on the following considerations: (1) According to Dr. Brill, potash glass made in the West during Medieval times usually contains 3-5% MgO or even more. In comparison, potash glass excavated in China contains very low MgO: usually less than 1%. In China, natural saltpeter

Figure 7. *Pieces of glass garment, similar to G52 and G53.*

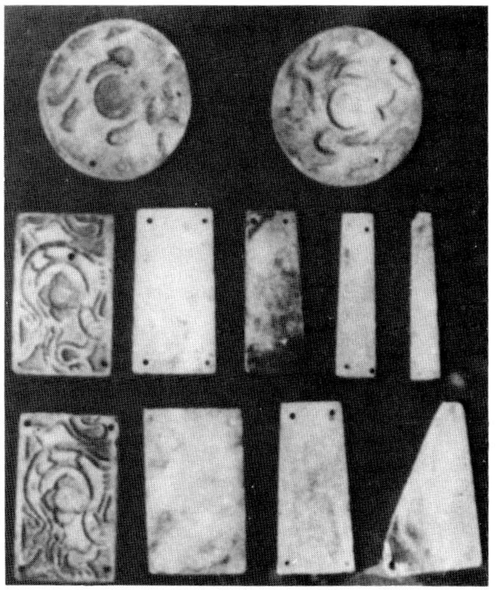

Figure 8. *Pieces of glass garment, similar to G52 and G53.*

Figure 9. *Decoration on piece of glass garment.*

Figure 10. *Decoration on piece of glass garment.*

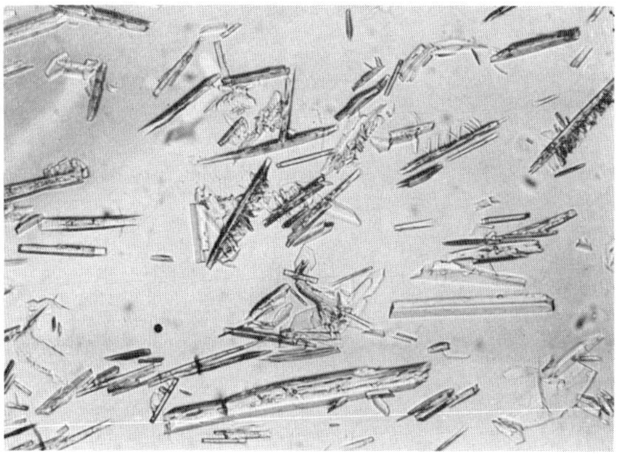

Figure 11. *Photomicrograph of barium disilicate crystals.*

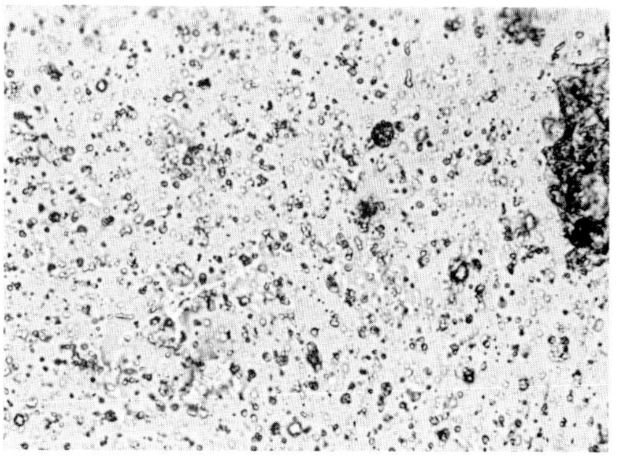

Figure 12. *Photomicrograph of G2 (~500X).*

(KNO_3) was readily available and was commonly used in alchemy or for making gunpowder. Natural saltpeter is considered the most probable source of potash for making glass in ancient China.[9] (2) Most of the blue potash glass has a high MnO content compared with the green potash glass. This indicates that the cobalt was of the high MnO type, which is characteristic of Chinese cobalt ores. A dark blue, barrel-shaped glass bead excavated at Kausambi, India, and dated 100-200 A.D., has a similar K_2O-SiO_2 composition. It has very low Na_2O, low CaO, moderate Al_2O_3 and low MgO. It is colored by cobalt which is accompanied by 1.48% MnO.[10]

Besides blue and green potash glass, a few red opaque potash glass beads have been excavated from a 1st-3rd century tomb in Guangxi Province. This is important, because the chemical composition of the copper red potash glass is quite different from those of the blue and green potash glasses. Its MgO content is as high as 3%. The manufacture of copper-red glazes by Chinese potters began during the 7th-10th century,[11] which is very late compared with the Middle East. Copper-red glass beads seem very rare in China; up to the present, only a few have been excavated, coming from tombs of the 1st-3rd centuries. On the other hand, they are very common in the Middle East. The provenance of this bead remains a question.

In addition to K_2O-SiO_2 glasses, some K_2O-CaO-SiO_2 glass relics have also been excavated in China. Figure 20 illustrates a bead from a 5th-3rd century B.C. tomb at Changsha, Hunan Province (no. G32, Table 1.) Figure 21 shows beads from a 5th-3rd century B.C. tomb in Sichuan Province (no. G22, Table 1.) and Figure 22 a hair pin from a 14th-century tomb at Yangzhou, Jiangsu Province (no. G68, Table 1).

Notes

1. Yang Boda, *Journal of the Palace Museum,* v. 2, no. 14, 1980, pp. 14-66.
2. Jiang Xuanyi, unpublished "Notes on Early Glass Excavated in China."
3. Zhang Fukang, Cheng Zhuhai, Zhang Zhigang, "An Investigation of Ancient Chinese *Liuli,*" *Journal of the Chinese Silicate Society,* v. 11, no. 1, 1983, pp. 67-76.
4. Robert H. Brill, Stephen S. C. Tong, and Zhang Fukang, "The Chemical Composition of a Faience Bead from China," *Journal of Glass Studies,* 1989, v. 31, pp. 11-15.
5. Zhang Fukang, "Origin and Development of Early Chinese Glass," *Archaeometry of Glass, 14th International Congress on Glass, New Delhi, 1986,* Calcutta, Indian Ceramic Society, Section II, 1987, pp. 25-28.
6. Gao Zhixi, "On Chinese Glasses of the 8th-3rd Centuries B.C.," *Wenwu,* no. 12, 1985.
7. See also this volume, Chapter 4.
8. Zhang Fukang, "Persian Glazed Pottery Excavated in Yangzhou," *2nd International Conference on the Science and Technology of Ancient Chinese Pottery and Porcelain,* Beijing, 1985.
9. Shi Meiguang, He Ouli, Zhou Fuzheng, "Investigations of Some Chinese Potash Glasses Excavated in Han Dynasty Tombs," *ibid.* note 5, Section II, pp. 15-20.
10. Robert H. Brill, "Chemical Analyses of Some Early Indian Glasses," *ibid.* note 5, Section I, pp. 1-25.
11. Zhang Fukang, "A Study of the Chemical Composition of Changsha Ware," *ibid.* note 8, Beijing, 1985.

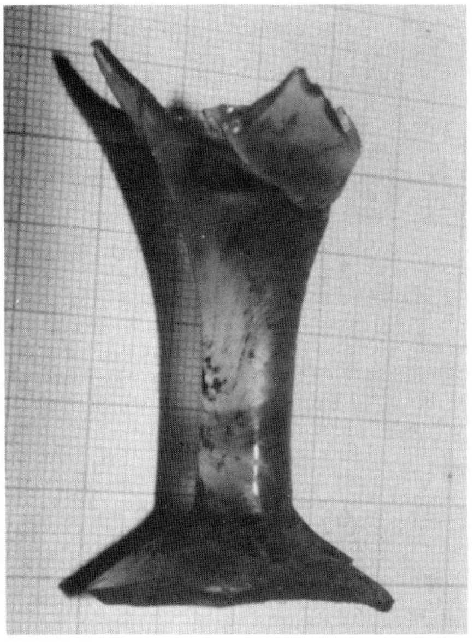

Figure 13. *Stem of wineglass. [Probably European.] (Soda-lime glass.)*

Figure 14. *Base of vessel. (Soda-lime glass.)*

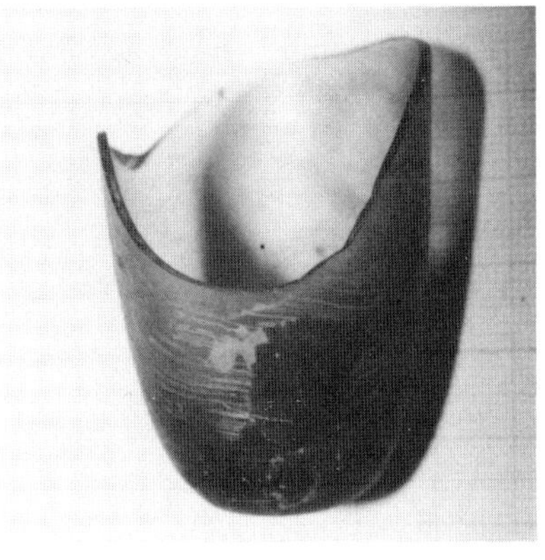

Figure 15. *Vessel fragment. (Soda-lime glass.)*

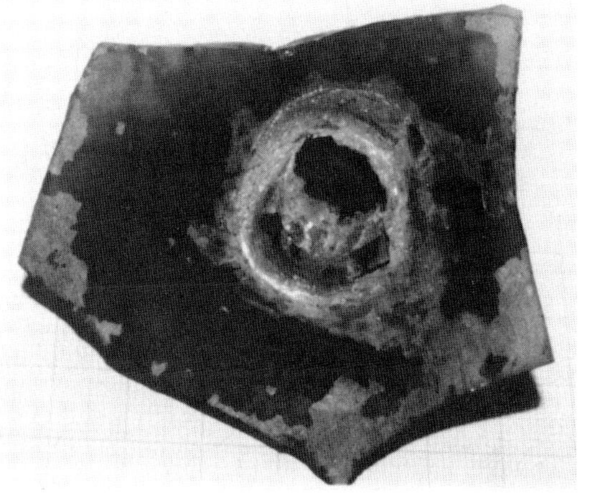

Figure 16. *Base of vessel. (Soda-lime glass.)*

Figure 17. *Fragment. (Soda-lime glass.)*

Figure 18. *Bead. (Soda-lime glass.)*

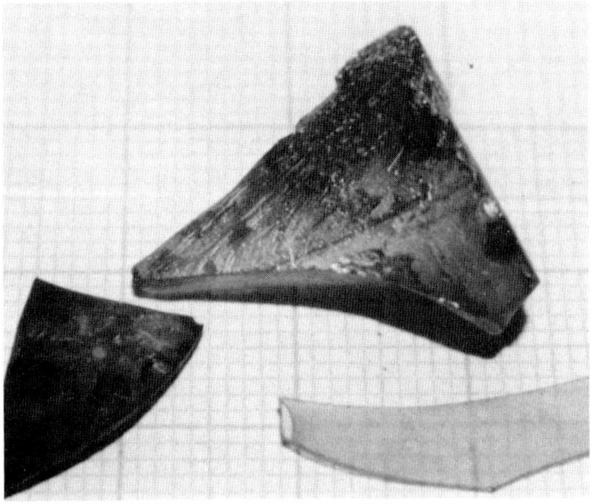

Figure 19. *Vessel fragments. (Soda-lime glass.)*

Figure 20. *Eye bead,* G32. *(Potash-lime glass.)*

Figure 21. *Eye beads,* G22. *(Potash-lime glass.)*

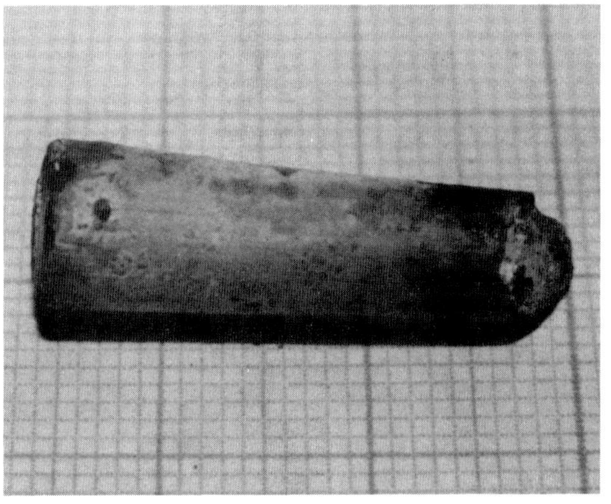

Figure 22. *Hair pin,* G68. *(Potash-lime glass.)*

Table 1

Analyses of Early Chinese Glasses

	G1	G6	*G7*	*G14*	*G8*	*G8W*	*G21*	*G41**	*G51*
SiO_2	90	94.0	94.11	major	32.6		56.53	38.77	18.02
Na_2O	1.2	0.3	0.44	0.64		0.07	7.42	1.75	0.03
CaO	0.4	0.4	0.18	0.35		0.52	0.70	0.56	1.65
K_2O	3.4	0.3	1.19	1.30		0.39	0.06	0.22	.01
MgO	0.3	0.2	0.06	0.15		0.04	1.67	0.11	0.01
Al_2O_3	0.3	0.7					0.69	2.30	0.08
Fe_2O_3	0.2	0.4		0.33		0.09	0.14	0.26	0.44
TiO_2					0.01	0.01	0.01	0.01	0.01
MnO	0.03					0.02	0.01	0.004	0.02
CuO	1.6	0.8		1.20		0.26	0.20	0.05	0.003
PbO				0.23	41.14	55.72	21.30	33.92	62.39
BaO					13.57	0.33	10.95	20.05	0.21
P_2O_5						5.39			4.38
L.O.I.		1.7				15.93			10.32
Sum.					86.98	78.77			97.57

*G41 contains CoO 0.04%

	*G52**	*G52***	*G53*	*G9*	*G10*	*G32*	*G22*	*G68*
SiO_2	34.83	22.56	36.15	27.88		73.73	70.86	50.98
Na_2O	2.16	0.06	2.11	0.13	0.21	0.50	0.89	0.64
CaO	0.36	1.50	0.25	0.22	0.33	4.51	5.35	6.15
K_2O	0.08	0.01	0.05	0.53	0.61	12.34	14.34	12.12
MgO	0.01	0.01	0.02	0.04	0.07	0.89	0.84	0.26
Al_2O_3	0.18	0.17	0.04	0.32	0.44	2.87	2.70	0.60
Fe_2O_3		0.48		0.20	1.77	1.42	1.26	0.23
TiO_2	0.01	0.01	0.01			0.26	0.22	
MnO	0.01	0.02	0.01			0.08	0.11	0.07
CuO	0.03	0.007	2.96			0.40	2.63	1.61
PbO	42.98	58.10	39.11	66.86	67.83	0.05	0.76	
BaO	17.38	1.55	20.01					
P_2O_5		4.96						
L.O.I.	1.03	11.10	1.52					
Sum.	99.03	100.53	99.28			99.28	98.94	

* G52 - Core.

** G52 - Weathered Crust.

CHAPTER 16

A COMPILATION OF LEAD ISOTOPE RATIOS OF SOME ORES FROM CHINA PUBLISHED BY CHEN YUWEI, MAO CUNXIAO, and ZHU BINGQUAN

ROBERT H. BRILL AND MARGERY CHEN

The Corning Museum of Glass,
Corning, N.Y.

IN 1980 Chen Yuwei, Mao Cunxiao, and Zhu Bingquan of the Institute of Geochemistry in Guiyang, published lead-isotope ratios for numerous galenas and other minerals and ores from various locations in China. That publication, in Chinese, also contained lengthy discussions of the geochemical aspects of the authors' findings.[1]

Great interest is now developing in lead-isotope studies of ancient Chinese glasses, bronzes, and other materials. These studies are undertaken with a view towards learning which objects might have been made in the same regions as one another, as opposed to which were made in different regions, and, under the most favorable circumstances, identifying those regions. The value of the technique is well evidenced in three papers included in this volume. (See Chapters 5, 6, and 14.) However, in order to gain the maximum benefit from analyses of ancient objects, it is necessary to have a library of lead-isotope ratios for mines or mining regions in China, mines which are believed to have been worked in ancient times. Only then is it possible to look for isotopic matches among the objects and ores and thereby draw inferences as to the possible geographical origins of the artifacts. Until recently, such data have been hard to come by. Therefore, the publication of Chen, Mao, and Zhu, although conducted for geological and geochemical reasons, could also prove of great value for those investigating the history of glassmaking — and other technologies — in ancient China.

In order to facilitate these investigations, one of the authors (R.H.B.) compiled the data and, along with help of the other author (M.C.), attempted to locate the sources of the samples on maps of China. For the most part, this has been successful but some of the locations remain, only for the present it is hoped, ambiguous.

In the tables which follow, the 121 ores are arranged alphabetically according to sample prefixes designating provinces or autonomous regions of China. A key appears below. The samples are numbered serially within each provincial group with numbers corresponding to the sample numbers in the original publication. The assumed locations are listed in both the Pinyin and Wade-Giles systems. Also given in each entry is the $^{207}Pb/^{206}Pb$ ratio reported in the original publication.

On the maps, sample locations are marked as circles or squares with their numbers. Circles indicate locations which the compiler and translator believe are accurate while squares indicate sample locations which are uncertain.

The shading of the location circles and squares gives some idea of the isotopic range within which the $^{207}Pb/^{206}Pb$ ratio falls for each sample. Open circles and squares designate low ratios (< 0.830), and completely shaded markers designate high ratios (> 0.870). Three intermediate ranges are designated by partial shading.

The compiler (R.H.B.) regrets that there has not been time to verify the map locations with the Chinese authors, and therefore takes full responsibility for any errors in locations on the map, for it is he who placed the markers. Approximately a dozen of the entries in the original publication are either ambiguous or do not appear on maps available to the compiler. Also, because he does not know Chinese, there may be peculiar inconsistencies in spelling or misunderstandings of other sorts which will be immediately obvious to some readers, but that should not detract too much from the usefulness of the data. Along with these apologies, also comes the hope that this effort will be of some help to scientists and archaeologists working in this extraordinarily promising area of archaeological chemistry.

References

1. The data in the accompanying table were published by Chen Yuwei, Mao Cunxiao, and Zhu Bingquan in "Lead Isotope Composition and Genesis of Phanerozoic Metal Deposits in China," *Geochimica*, vol. 3, no. 9, September, 1980, pp. 215-229. (In Chinese.)

Key to Symbols on Map

Low ratios	(9)	<0.830
Intermediate ratios	(23)	0.830-0.840
Mid-range ratios	(65)	0.840-0.860
Intermediate ratios	(16)	0.860-0.870
High ratios	(8)	>0.870
	(n = 121)	

Square indicates precise location unknown.

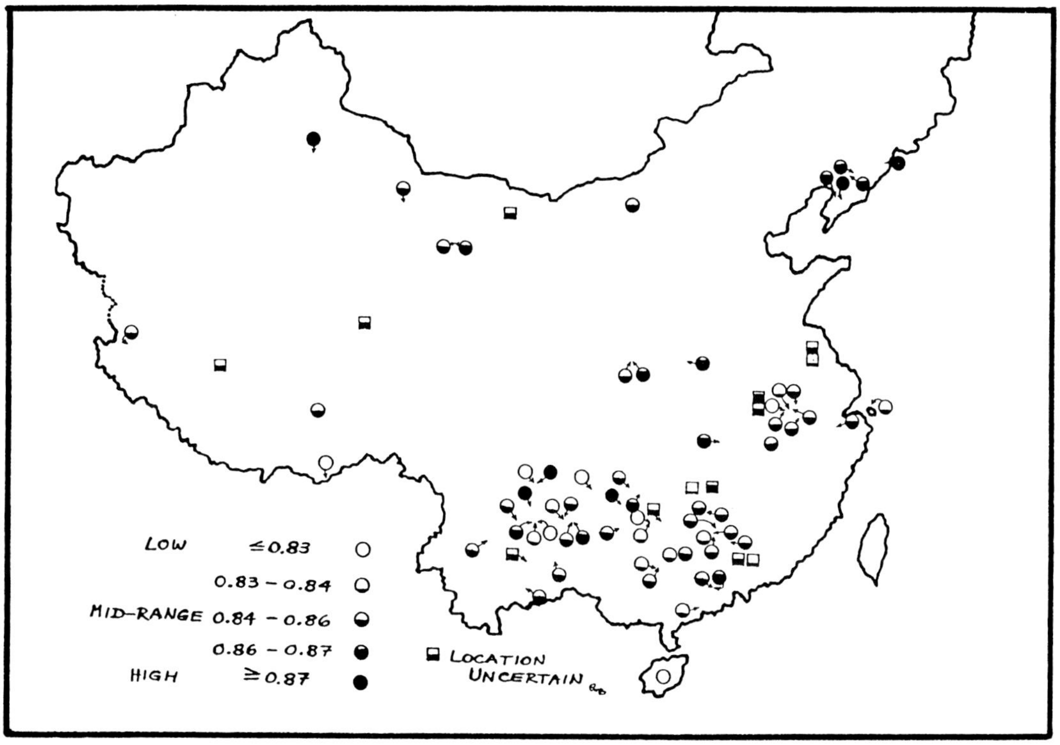
LOW ≤0.83
0.83 – 0.84
MID-RANGE 0.84 – 0.86
0.86 – 0.87
HIGH ≥0.87
LOCATION UNCERTAIN

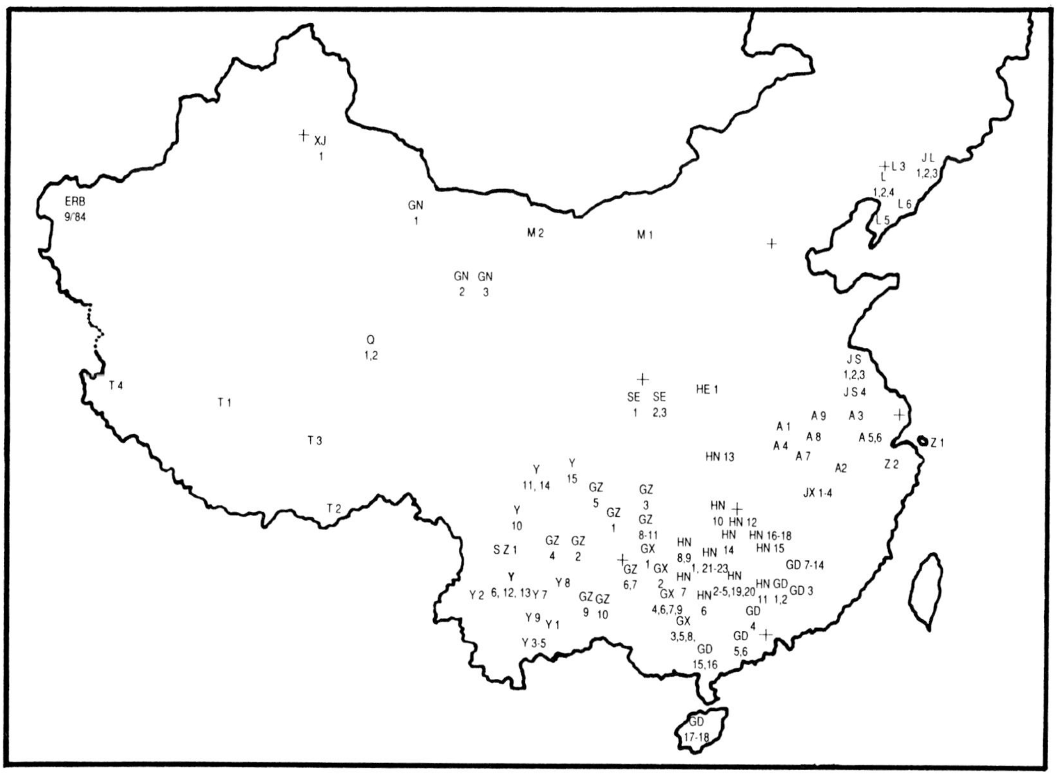
XJ 1
ERB 9/84
GN 1
M 2
M 1
L 3
JL 1,2,3
L 1,2,4
L 6
L 5
GN 2
GN 3
Q 1,2
JS 1,2,3
JS 4
T 4
T 1
HE 1
SE 1
SE 2,3
T 3
T 2
A 9
A 3
A 1
A 8
A 5,6
A 4
A 7
A2
Z 1
Z 2
HN 13
JX 1-4
GD 17-18

Provinces and Prefixes

A	ANHUI (Anhwei)	L	LIAONING (Liaoning)
F	FUJIAN (Fukien)	Q	QINGHAI (Tsinghai)
GN	GANSU (Kansu)	SX	SHANXI (Shansi)
GD	GUANGDONG (Kwangtung)	SH	SHANDONG (Shantung)
GZ	GUIZHOU (Kweichow)	SE	SHAANXI (Shensi)
HO	HEBEI (Hopeh)	SZ	SICHUAN (Szechwan)
HL	HEILONGJIANG (Heilungkiang)	Y	YUNNAN (Yunnan)
HE	HENAN (Honan)	Z	ZHEJIANG (Chekiang)
HU	HUBEI (Hupeh)	GX	GUANGXI ZHUANG (Kwangsi)
HN	HUNAN (Hunan)	M	NE MENG GU (Inner Mongolia)
JX	JIANGXI (Kiangsi)	N	NINGXIA HUI (Ningsia)
JS	JIANGSU (Kiangsu)	XJ	XINJIANG UIGHUR (Sinkiang)
JL	JILIN (Kirin)	T	XIZANG (Tibet)

Notes

1. The data are listed here in approximately the same order in which they appear in the original Chinese publication and are separated according to provinces and autonomous regions.
2. Numbers have been assigned. These have prefixes corresponding to the provinces and are numbered serially within the provinces. The sample numbers from the original Chinese publication are also given.
3. The Wade-Giles entries are the translations as supplied by Mrs. Chen.
4. An asterisk (*) indicates sites which are believed to have been located accurately.

PROVINCE	NO. ON MAP	CHINESE SAMPLE NO.	LOCATION (PINYIN)	LOCATION (WADE-GILES)	$^{207}Pb/^{206}Pb$ RATIO	
ANHUI (ANHWEI)	A1	DH001	Dishuiya	Ti-shui-yia	0.8611	
	A2	V001	Huang Shan	Huang Shan	0.8537	*
	A3	V002	Hengshan	Heng Shan	0.8588	*
	A4	V003	Xueshuifu	Hsueh Shui Fu	0.8519	
	A5	1	Qingyang, Wuzhang Miao (temple)	Ch'ing-yang, Wu-Chang Miao (temple)	0.8532	*
	A6	1	Same	Same	0.8512	*
	A7	5	Guiji	Kuei-Ch'ih, Hsiao-ho-lu	0.8459	*
	A8	6	Same	Same	0.8234	*
	A9	3	Tongling	T'ung-ling, Feng-huang Shan	0.8341	*
ZHEJIANG (Chekiang)	Z1	299	Daishan Xian, Chencaizhun	Tai-shan Hsien, Ch'en-ts'ai-chun	0.8373	*
	Z2	1	Zhuji (or Zhuqi?) Dongyang Shan (Lead and zinc mine)	Chu-chi, Tung-yang Shan (Lead and zinc mine)	0.8512	*
JIANGXI (Kiangsi)	JX1	G18	Jiugiang, Wu Shan (Lead and zinc mine) (near Jinqdezhen)	Chiu-chiang, WuShan (Lead and zinc mine)	0.8531	*

	JX2	G19	Same	Same	0.8515	*
	JX3	G9	Jiujiang, Zhengmen Shan (Lead and zinc mine)	Chiu-chiang, Cheng-men Shan (Lead and zine mine)	0.8505	*
	JX4	4	Same	Same	0.8554	*
GUANGDONG (Kwangtung)	GD1	Pb-01	Guidong (Many metal mines)	Kwei-tung (Many metal mines)	0.8550	
	GD2	Pb 07	Dabao Shan (Many metal mines)	Ta-pao Shan (Many metal mines)	0.8421	
	GD3	D (?)	Same	Same	0.8352	
	D4	Pb-02	Shaoqing, Dajiangbing (near Foshan)	Chao-ch'ing, Ta-chiang-ping	0.8501	*
	GD5	Pb-06	Same	Same	0.8609	*
	GD6	Pb-08	Same	Same	0.8683	*
	GD7	5	Renhua, Fankuo (Lead and zinc mine)	Jen-hua, Fan-k'uo, Chin-ksing-ling (Lead and zinc mine)	0.8526	*
	GD8	6	Same	Same	0.8558	*
	GD9	V-2	Same	Same	0.8507	*
	GD10	F-42	Renhua, Fankuo	Jen-hua, Fan-k'uo Chung-shang-ni-pen-chi	0.8567	*
	GD11	V-28	Renhua, Fankuo (Lead and zinc mine)	Jen-hua, Fan-k'uo Shih-tzu-ling. (There is a mountain with this name in Kiangsi.) (Lead and zinc mine)	0.8426	*

	GD12	V-121	Same	Same	0.8560	*
	GD13	V-132	Same	Same	0.8529	*
	GD14	V-70	Same	Same	0.8482	*
	GD15	LK-15	Yangchun (Tungsten mine)	Yang-ch'un, Hsiao-nan Shan (Tungsten mine)	0.8345	*
	GD16	E-203	Same	Same	0.8317	*
	GD17	S-76-30	See Hainan			
	GD18	S-76-160	See Hainan			
HUNAN (Hunan)	HN1	2	Chenxian, Dongbo, Hengshan	Ch'en Hsien, Tung-po, Heng Shan	0.8464	*
	HN2	10	Same	Same	0.8380	*
	HN3	4	Same	Same	0.8320	*
	HN4	8	Same	Same	0.8395	*
	HN5	D-23	Same	Same	0.8377	*
	HN6	2	Jiangyong (Lead and zinc mine)	Chiang-yung, Tung Shan (Lead and zinc mine)	0.8414	*
	HN7	3	Same	Same	0.8334	*
	HN8	38	Taolin (Mountain or island?) (Lead or zinc mine) (There are 3 Taolins in Hunan.)	T'ao-lin Chien Chen-tan-ti-tseng (Lead and zinc mine)	0.8533	* (?)
	HN9	49	Same	Same	0.8501	* (?)

HN10	TW-24	Chai Shan (Lead and zinc mine)	Chai Shan (Lead and zinc mine)	0.8321	
HN11	W-7	Huangshabing (Lead and zinc mine) (Huangshabao has been located.)	Huang-sha-ping (Lead and zinc mine) (Huang-sha-pao has been located.)	0.8472	*
HN12	T-1	Huadan (Lead and zinc mine) (There is a Huat'an in Sichuan.)	Hua-tan (Lead and zinc mine) (There is Hua-t'an in Szechwan.)	0.8631	* (?)
HN13	Pb-20	Linxiang	Lin-hsiang, Tao-lin	0.8617	*
HN14	Pb-09	Shaodong (Xian?)	Shao-tung, Mao-yen-ti Shao-tung Hsien	0.8564	*
HN15	Pb-03	Qitian Ling	Ch'i-t'ien Ling	0.8523	*
HN16	Pb-19	Shuikoushan (Lead and zinc mine)	Shui-kuo Shan (Lead and zinc mine)	0.8442	*
HN17	4	Same	Same	0.8401	*
HN18	Pb-08	Same	Same	0.8528	*
HN19	4	Chenxian (Lead and zinc mine)	Chen Hsien, Tung-po (Lead and zinc mine) (Eastern Lake?)	0.8370	*
HN20	8	Same	Same	0.8380	*
HN21	5	Same	Same	0.8523	*
HN22	4	Chenxian, Dongbo (Sichuan?)	Chen Hsien, Tung-po, (Szu-chu-yuan)	0.8414	*
HN23	11	Same	Same	0.8399	*

HENAN ((Honan)	HE1	1	Zhuanbei, Daheju Liushanyai	Chuan-pe, Ta-ho Chu, Liu Shan Yai	0.8600	*
YUNNAN (Yunnan)	Y1	Pb-06	Luoping, Fuluo Zhang (Factory)	Lo-p'ing, Fu-le Chang (Factory)	0.8488	*
	Y2	65009	Yaoan, Lao jian zi	Yao-an, Lao-chian-tzu	0.8555	*
	Y3	L-38	Gejiu (Tin mine)	Ko-chiu (Tin mine)	0.8420	*
	Y4	L-59	Same	Same	0.8502	*
	Y5	L-64	Same	Same	0.8463	*
	Y6	I-101	Huize (Mineral Mountain)	Hui-tse (Mineral Mountain)	0.8532	*
	Y7	I-109	Same	Same	0.8348	*
	Y8	I-102	Huize (Five Metals Factory. The five metals are: gold, silver, copper, tin, and iron.)	Hui-tze (Five Metals Factory. The five metals are: gold, silver, copper, tin, and iron.)	0.8039	*
	Y9	I-103	Huize, Yiliang fault plane (There are two Yiliangs.)	Hui-tze, I-liang-mao-ping (There are two I-liangs.)	0.8451	*
	Y10	I-104	Huize, Qiaojia maozhu	Hui-tze, Chiao-Chia-mao-chu	0.8599	*
	Y11	I-106	Huize, Lanying Mine (Blue silver mine)	Hui-tze, Lan-ying Mine (Blue silver mine)	0.8257	
	Y12	I-107	Huize, Jilingzhang (Factory)	Hui-tze, Chi-ling Chang (Factory)	0.8461	

	Y13	I-108	Same	Same	0.8433	
	Y14	I-105	Huize, Yongshan (Corundum factory)	Hui-tze, Yung Shan (Corundum factory)	0.7624	*
	Y15	C-8	Huize (Lead and zinc mine)	Hui-tze (Lead and zinc mine)	0.8597	*
GUANGXI ZHUANG (Kwangsi)	GX1	4	Linqi (?), Fuzizhong (Lead and zinc mine)	Lin(?)-chi, Fu-tzu-chung (Lead and zinc mine)	0.8267	*
	GX2	2	Same	Same	0.8310	*
	GX3	Pb-30	Dazhang, Bali (Many metal mines)	Ta-chang, Pa-li (Many metal mines)	0.8525	*
	GX4	Pb-34	Dazhang, Fulongzhang (Many metal mines)	Ta-chang, Fu-lung-chang (Many metal mines)	0.8360	*
	GX5	W-047	Dazhang, Bali (Many metal mines)	Ta-chang, Pa-li (Many metal mines)	0.8440	*
	GX6	W-250	Dazhang, Zhangbo	Ta-chang, Chang-po Mine	0.8362	*
	GX7	W-027	Dazhang, Bali	Ta-chang, Pa-li Mine	0.8384	*
	GX8	W-739	Dazhang, Longxianggai (Lead and zinc mine)	Ta-chang, Lung-hsiang-kai (Lead and zinc mine)	0.8441	*
	GX9	W-729	Same	Same	0.8387	*
GUIZHOU (Kweichow)	GZ1	Pb-24	Youjing (Baini) (A county?) (Lead and zinc mine)	Yu-ch'ing (Pai Ni) (Yu-ching is a county?) (Lead and zinc mine)	0.8766	*

	GZ2	26	Shuicheng (Xian?), Shanshuling (Lead and zinc mine)	Shui-cheng (Hsien?) Shan-shu-ling (Lead and zinc mine)	0.8422	*
	GZ3	Pb-21	Sinan (Lead and zinc mine)	Ssu-nan (Lead and zinc mine)	0.8507	*
	GZ4	23	Shuicheng (Xian?) Shanshuling (Lead and zinc mine)	Shui-cheng (Hsien?) Shan-shu-ling (Lead and zinc mine)	0.8375	*
	GZ5	Pb-17	Renhuai (Lead and zinc mine)	Jen-huai (Lead and zinc mine)	0.8205	*
	GZ6	Pb-27	Puan, Dingdoushan	P'u-an, Ting-tou Shan	0.8535	*
	GZ7	Pb-28	Puan, Lumaobing	P'u-an, Lu-mao-ping	0.8581	*
	GZ8	2	Wanshan Pugaozhang (Mercury mine)	Wan Shan, Pu-kao-chang (Mercury mine)	0.8590	*
	GZ9	3	Wanshan Huachu Shan (?) (Mercury mine)	Wan Shan, Hua-ch'u Shan (Mercury mine)	0.8479	*
	GZ10	4	Same	Wan Shan, Hua-ch'u Shan (Mercury mine)	0.8613	*
	GZ11	5	Wanshan (Mercury mine)	Wan Shan, Yen-sou-ping (Mercury mine)	0.8644	*
SICHUAN (Szechwan)	SZ1	3(?)	Huidong	Hui-tung	0.8407	*
SHAANXI (Shensi)	SE1	Xzy-1	Zhashui (Xian?) Yindongzi (Lead and zinc mine)	Cha-shui (Hsien?), Uin-tung-tzu (Lead and zinc mine)	0.8507	*
	SE2	Xzy-2	Same	Same	0.8643	*
	SE3	Xzy-3	Same	Same	0.8641	*

GANSU (Kansu)	GN1	Pb-12	Huaniushan (In northwest Gansu)	Hua-niu-Shan (In northwest Gansu)	0.8452	*
	GN2	Pb-10	Qilian Shan	Chi'-lien Shan (Blue copper)	0.8501	*
	GN3	Pb-11	Qilian Shan Xiaotie Shan (Lead and zinc mine)	Chi'-lien Shan (Hsia t'ieh Shan) (Lead and zinc mine)	0.8688	*
QINGHAI (Tsinghai)	Q1	Pb-16	Beisi Shan (?) (North Tin Mountain) (Lead and zinc mine)	Pe-si (?) Shan (North Tin Mountain) (Lead and zinc mine)	0.8575	*
	Q2	0576	Same	Same	0.8525	*
XIZANG (Tibet)	T1	73-3	Kangdin (or Gangding) Cheng, Dong Shihying Mai (?) (East Quartz Mountain) (There is a Kangding in Sichuan.)	K'ang-ting Ch'eng (City), Tung Shih-ying Mai (East Quartz Mountain)	0.8437	*(?)
	T2	73-546	Chayu Shihying Mai	Ch'a-yu Hsien, Shih-ying Mai	0.8281	*
	T3	76-786	North of Lhasa	North of La-sa Shih	0.8472	*
	T4	76-118	Ali Diqu (Ali Region) Shiquanhe	A-li Ti-Ch'u Shih-chuan-ho (River)	0.8454	*
LIAONING (Liaoning)	L1	Pb-38	Fengcheng, Jingzhengzi (Lead and zinc mine)	Fung-ch'eng, Ching-cheng-tzu (Lead and zinc mine)	0.8791	*
	L2	7	Same	Same	0.8707	*
	L3	10	Same	Same	0.8669	*

	L4	Pb-25	Same	Same	0.8828	*
	L5	Pb-41	Gaibing, Fulungquan (?)	Kai-ping, Fu-lung-chuan (?)	0.8649	*
	L6	No. 3	Lianshanguan Baishuishi (?)	Lien-shan Kuan, Pai-shui Shi	0.8640	*
JILIN (Kirin)	JL1	Pb-36	Jian, Hanwu, Aotouri (?) (Lead and zinc mine)	Chi-an, Han-wu Ao-tou-ji (Lead and zinc mine)	0.8730	*
	JL2	G13C	Jian, Hanwu (Lead and zinc mine)	Chi-an, Han-wu (Lead and zinc mine)	0.8749	*
	JL3	G13B	Jian, Hanwu (Lead and zinc mine)	Chi-an, Han-wu (Lead and zinc mine)	0.8846	*
NE MENG GU (Inner Mongolia)	M1	Pb-23	Daqing Shan (Mountain range above Huhehot)	Ta-ching Shan	0.8461	*
	M2	Pb-14	Suosuoqing (?) (Western part of Inner Mongolia)	Suo-suo Tsing (?) (Western part of Inner Mongolia)	0.8488	* (?)
XINJIANG UIGHUR (Sinkiang)	XJ1	Pb-15	Kalanguguole (Near Qichuanhu)	K'a-lan-ku	0.8769	*
HAINAN ISLAND (Guangdong)	GD17	S-76-30	Shilu, Tiandu (Iron mine)	Shih-lu, T'ien-tu Mine (Iron Mine)	0.8195	*
	GD18	S-76-160	Same	Shih-lu, T'ien-du Mine (Iron mine)	0.8269	*

JIANGSU (Kiangsu)	JS1	L-19	Ningwu Iron Mine	Ning-wu Iron Mine	0.8489
	JS2	L-28	Same	Same	0.8579
	JS3	L-21	Ningwu Iron Mine, Anshanyan Yanshijian (There are Anshans in 3 other provinces.)	Ning-wu Iron Mine, An-shan-yen, Yen-shih Chien	0.8459
	JS4	L-22	Same	Same	0.8307

CHAPTER 17

A HAN DYNASTY GLASS SPEARHEAD FROM CHANGSHA

XIONG ZHUANXIN

The Hunan Province Museum,
Changsha, Hunan

IN 1956 a glass spearhead (Figs. 1 and 2) was found in a Western Han Dynasty tomb at Sha Hu Qiao (The Sandy Lake Bridge), a northern suburb of Changsha in Hunan Province.[1] The glass is a bluish-aqua transparent color. We also found in the tomb a large quantity of other glass objects including *bi* disks, beads, tubes, sword *shou* (pommel decorations), sword *er* (ring-shaped decorations in the middle of the scabbard), sword scabbard terminals, and seals. The objects date from the Warring States Period to the Western Han Dynasty. Among the glass found, the glass spearhead is the only weapon-shaped object. It may play some important role in the study of early Chinese glass, because to our knowledge no other object of this type made of glass has ever been found.

The spearhead was unearthed in a clay tomb pit, 3.04 m long and 1.52 m wide. The coffin was almost totally decayed; only the bottom boards and two pieces of supporting wood remained. The glass spearhead was found on the remains of one of the pieces of wood, that on the northern side, and pointed towards the northwest. More than 30 funeral objects, including bronze lamps, a bronze cross-bow, pottery jars, an iron knife, and pottery Wuzhu coins were unearthed along with the well-preserved glass spearhead.

The object consists of a shaft and blade formed as a single piece. The total length is 18.8 cm, the blade being 9 cm and having no sharp tip or sharp cutting edge. There is a ridge along the center line of the body, and to each side there is a blood channel. The greatest width between the two edges is 2.6 cm. The shaft is circular in cross section, being thicker in diameter (1.2 cm) near the body and tapering toward the terminal (0.6 cm). The shaft is 9.8 cm in length with a pommel-like ball in the mid section. The surfaces bear obvious traces of grinding and polishing, and the glass contains a few small bubbles.

It was reported that the tomb and spearhead are of the late Western Han Dynasty, but, for two reasons, the author believes the date to be middle Western Han. The Wuzhu coins unearthed with the spear are characteristic of the reign of Emperor Wu in Yuan Shou (113 B.C.), and they differ from those made during the reign of Emperor Xuan (73-49 B.C.). On the coins found in the tomb (which are about 2 cm in diameter) the form of the strokes in certain Chinese characters are more like those on the earlier coins. All the pottery also belongs to the middle Western Han Dynasty. Consequently, the glass spearhead was used as a funeral object about 2100 years ago.

In ancient times, spearheads were commonly used in both the East and the West. However, they were always made of stone or metal. Because this piece does not have a sharp tip or blade edges, and because glass is so fragile, this spearhead could not have been a weapon. It must have been only a ceremonial object or an ornamental work of art.

The shape of the glass spearhead poses a problem in that it is different from that of bronze spears excavated in the tombs of the Warring States Period (475-221 B.C.) in Changsha and elsewhere in China. We have never seen a shaft with a ball in the mid section among Chinese bronze spears of that date. This suggests that the glass spearhead may not have been made in China.

The specific gravity of the spearhead has been determined by Shi Meiguang of The Research Institute of Building Materials and by Gao Zhixi of The Museum of Hunan Province. The weight of the object in air is 37.35 g, and in water 21.5 g. The resulting specific gravity is 2.47. {Editor's Note: There may be an error in transcription. The reported weights calculate out to a specific gravity of 2.36.} Chemical analyses show that most glass *bi* and beads excavated near Changsha belong to the lead-barium system and such glasses have specific gravities greater than 3. Thus it is clear that the spearhead is not made of a lead-containing glass.[2] This glass most probably belongs to the soda-lime {or potash-lime} system. It is therefore possible that the glass was imported from the West. It is known that some eye beads found in China were imported. As early as the 5th-4th centuries B.C. trade between the Eastern and the Western Worlds is thought to have included glass.

Changsha is located south of the Yangtze River. As early as the Warring States Period and Western Han Period, Changsha had become a very important city. Changsha, the capital of Changsha State, had the advantage of having convenient transportation along the Xiang River and its branches. Travelers could reach Guangdong and Guangxi across the southern mountain ranges by means of this river system. Penchang (Guangzhou) and Hepu were important ports of entry from foreign countries at that time. Goods from East and West Asia and from the Mediterranean came into China through these ports.

In *Shiji (A Record of History)* and *Hanzhi Liezuan (Trade and Commerce History)* we find the following record: "Panyu is also a commercial capital where pearls, jade, rhinoceros horn, tortoise shell, fruits, and clothes were traded {?}." In addition, western style dragonfly-eyed glass beads have been found in Western Han Dynasty tombs in Guangdong and Luobo Bay in Guangxi.

During the reign of Emperor Wu, according to records in *Hanshu Dilizi* (a Han Dynasty book about geography), commercial fleets led by an official representing Emperor Wu departed from the port of Hepu. They sailed across the southern sea to Southeast Asia and South Asia. There they obtained precious jewels and treasures from Roman merchants and brought them back to China. The Roman goods included glasswares, which were considered luxurious by the Chinese who traded silk and gold in return.

In the Han Dynasty book *Houhan Shu Xiyuzhuan (History of Turkestan)* is written: "Da Qin (Rome) is rich in gold, silver, rare jewelry and other odd treasures, night-gleamed *bi* {?}, bright moon pearl {?}, coral, and *liuli* (glass)."

In the Wei Dynasty book, *Wei Shu Xiyuzhuan (History of Turkestan)*, another description of Roman {?} glass is more vivid and complete. The colors of the glass are described as "red, white, black, green, blue, dark purple, light green, and purple." Glass cups and other objects which might have been made in ancient Rome have been excavated in the tomb of the prince of Zhongshan in Mancheng, (Hebei Province) and in tomb no. 2 in Ganquan, Hanquan (Jiangsu Province).

Both the records and the objects illustrate that western glass was imported into China after the middle Han Dynasty. Because the glass spearhead excavated in Changsha can be dated to the time glass was imported from the West, and because its shape is different from actual Chinese bronze weapons, it may be that the glass spearhead was imported from ancient Rome or Western Asia. Moreover, judging from the size of the tomb where the spearhead was found, and its elaborate contents, the occupant of the tomb must have been a person of wealth and high social rank. He might have been a noble of ancient Changsha, and a person of such rank could be expected to have owned valuable imported glass.

Notes

1. Li Chenguan, "The Excavation of Ancient Tombs in the Region of Shahu Bridge, Changsha," *Acta Archaeologica Sinica,* v. 3, 1987.
2. Robert H. Brill and John F. Wosinski, "Physical Properties of Early Chinese Glasses," this volume, Chapter 9.
3. Shi Meiguang, He Ouli, and Zhou Fuzheng, "An Investigation of Some Chinese Potash Glasses Excavated from Tombs of the Han Dynasty," *Journal of the Chinese Silicate Society,* v. 14, no. 3, 1986, pp. 307-313.

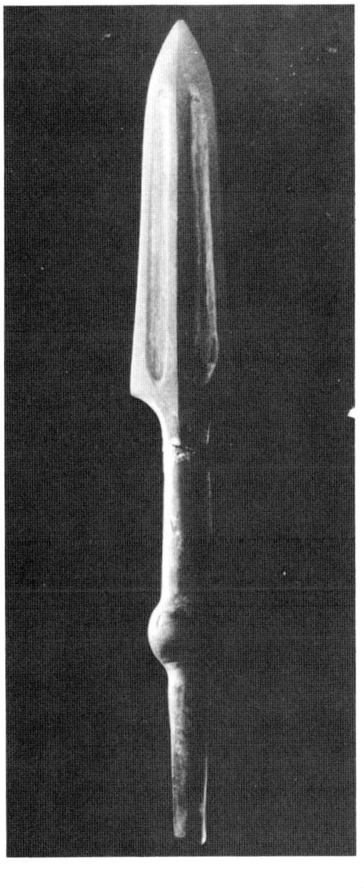

Figure 1. *Glass spearhead excavated at Changsha. Cast, aqua-colored glass, length 19.8 cm.*

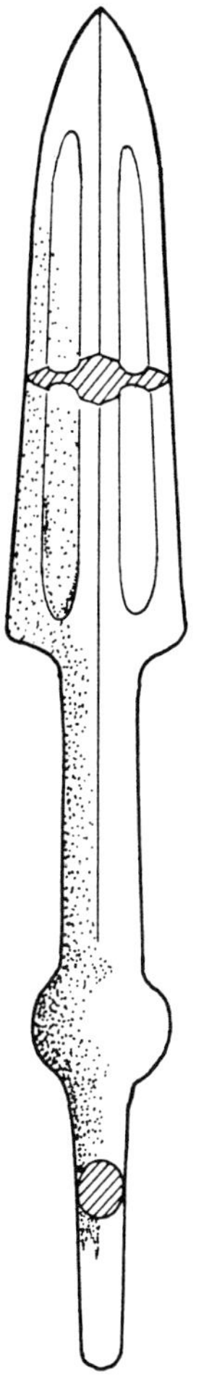

Figure 2. *Glass spearhead and sketch of coin.*

CHAPTER 18

A PRELIMINARY STUDY OF HAN DYNASTY GLASS IN GUANGXI

HUANG QISHAN

The Museum of Guangxi,
Zhuang Autonomous Region

UNTIL now, little has been published about glass from Guangxi, even though in recent years numerous glass objects have been found in the excavations of the Han Dynasty (206 B.C.-220 A.D.) tombs there. Some experts had studied the shapes of the objects, but they did not analyze their compositions. In many cases they concluded that the glass excavated in these tombs was imported from the West,[1] perhaps from the Roman Empire.[2] Other experts believed they show a similarity to ornaments made before the Qin Dynasty, and these experts contend that we cannot assume the beads were imported just because Guangxi happens to be situated on the land and water routes to western countries.[3] Therefore, it is important to consider whether all the glass objects excavated in Guangxi were imported or whether some were made in China.

In ancient Chinese literature, glass was called *liuli*[4] or *boli*.[5] To Chinese archaeologists, these mean "glassware" or "glazed ware."[6] The results of studies by scholars in China and abroad have proved that glass objects were made in China as early as the Spring and Autumn (770-476 B.C.) and Warring States (475-221 B.C.) Periods, although glass of these periods has not yet been found in Guangxi.

According to incomplete statistics (see Table 1), there are about 160 tombs at Hepu, Guixian, Zhaoping, Xingan, Wuzhou, Hexian, and Luijiang. Six kinds of glass objects, namely, beads, *bi* disks, tubes, ear spools, cups, and bowls have come from the tombs. Most are beads: up to 4,300 have been found. Generally speaking, there is at least one bead in each tomb. The largest number of beads found is 1,080 beads from the third Western Han Dynasty tomb at Tangpai, Hepu.

There are two kinds of beads, transparent and opaque. The colors of the transparent beads are bluish green, light bluish green, green, and lake-water blue. The opaque beads are sky-blue, bluish green, lake-water blue, green, black, white, moonlight-white, brick-red, and purplish-brown. The beads occur in three shapes, the overwhelming majority being the shape of beads used today on an abacus. The largest is 0.5 cm while the smallest 0.2 cm (Fig. 1). The second shape is prismatic, but only a few of these have been found. The third type has the shape of an olive. There are one or two air bubbles at the surface of some of the beads, and many have bubbles on the inside. Although they have been buried for more than 1,000 years, most of the beads are only lightly weathered and are firm and lustrous.

The other glass objects are five round, prismatic blue glass tubes, two blackish-green glass ear spools (Fig. 2), two blue bowls, one blackish-green disk (Fig. 3), and five glass cups, two of which are short-stemmed. One of the stemmed

Figure 1. *Glass beads of the western Han Dynasty.*

Figure 2. Glass ear spool of the Eastern Han Dynasty, Zhaoping. Analysis sample no. 6. (Lead-barium glass.)

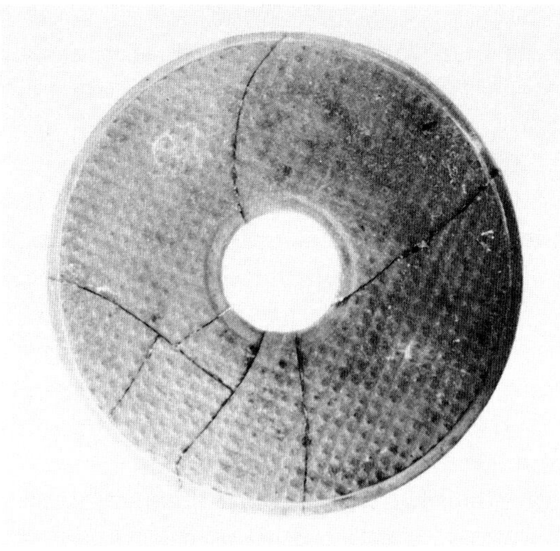

Figure 3. *Glass* bi *disk of the Western Han Dynasty.*

Figure 4. *Glass stemmed-cup with saucer of the Eastern Han Dynasty, Nandao.*

Figure 5. *Glass cup of the Eastern Han Dynasty.*

cups was excavated in the second tomb on Fengliu Hill, Guixian, in 1977. A blue glass cup, with a flat bottom, was excavated in the same tomb, but owing to the collapse of the coffin, the glass cups were crushed, and they cannot be restored. The chemical analyses of the two cups are similar: both are potash-silica glasses. The other nearly intact stemmed cup was excavated from the Eastern Han Dynasty tomb no. 3 at

Nandao Village, in the eastern suburb of Guixian. This type of cup, with a saucer, is quite rare. The mouth of the cup is decorated with a line and the side with two lines. The saucer, with its shallow rim and flat bottom, has a round depression in the center into which the cup fits. It is a transparent, light bluish-green glass, with ice-crack patterns. The cup is 8.2 cm high, with a rim diameter of 6.4 cm. The saucer is 5.2 cm in diameter.

In addition, two glass cups with flat bottoms were excavated from the Eastern Han Dynasty tomb at the railway station in Guixian, and one of these two cups is transparent. It has a wide-mouth, a rolled rim, and it is straight sided. The outside is decorated with two lines. It stands 3.2 cm high, with a rim diameter of 5.8 cm. This is also a potash-lime glass,[7] and it is now in the Museum of Chinese History. {Editor's Note: See Chapter 19.} The other cup from this tomb is blue opaque, with a wide mouth and straight sides. The outside is decorated with a line. It is 4 cm high with a rim diameter of 7.7 cm. The glass has not been analyzed (Fig. 5). It is similar to the green glass cup with a flat bottom mentioned above. Some experts consider that the two cups differ from the traditional glass of China and were therefore imported from the Near East.[8]

Shapes and Decoration. As was noted above, most of the glass beads found in Guangxi are abacus-shaped, and they are similar to the beads excavated in Central China which date before the Qin Dynasty. As a result, some experts believe they were made in China.[9] The *bi* disks and ear spools are typical Han Dynasty Chinese shapes and the trellis designs on the back of the glass disks are often seen in Han Dynasty bronzes and pottery. As to the glass cups, some experts believe they were imported,[10] but foreign experts have said that the line decorations are seldom seen on western glass. Therefore, it is quite probable that the cups were made in China.

Ancient Chinese Documents. Although there are no glassmaking references before the Qin Dynasty (221-207 B.C.), some clues about glass can be found in documents after the Han Dynasty. For example, Wang Chung, during the Eastern Han Dynasty, wrote that alchemists made *yangsui* by melting five kinds of mineral powders and that it was possible to make a fire with *yangsui* in the sun. Two pieces of very transparent glass lenses excavated in the tombs of the Caocao family at Boxian, Anhui Province, are the best archaeological evidence of such glass.[11]

Some other examples of early references follow. In the first year of Yuanding during the reign of Emperor Hanwu, "Zhaoxian Pavilion, with glass window panes, was built to the west of Ganquan Palace."[12] In the third year of Taichu, "When Wangfeng Terrace was built, there was a plant with white flowers on it." The Emperor Hanwu asked his subject to put the plant in a glass box and to give it to Dong Yan.[13] "Dong Yan often lived in a bedroom called Yanqin in which there was a screen decorated with purple glass."[14] At the beginning of the Han Dynasty, many officials made "seven glass trees" and presented them to Shanglin Park. Owing to the fact that much glass was produced, "Many of the windows and doors were paned with glass"[15] during the period of Emperor Hanwu.

Glass was so widely used in the Han Dynasty that most of the glass was probably made in China. At that time, Chinese glass was also exported. The glass hook and hexagonal glass rod of the Han Dynasty, presently in the National Museum in Tokyo, were probably imported from China.[16] An expert engaged in textual research on glass beads excavated in Malaysia believes that "most of the beads were imported from the Middle East, but some were imported from China, from the Christian Era or earlier."[17] Ge Hong, who made pills of immortality during the Jin Dynasty, wrote *Baopuzi neiwaibian,* in which he said, "The foreign crystal bowls were made from five kinds of mineral powders." Through textual research, some think that the crystal bowls were actually glass bowls.[18]

Chemical Analyses. Shi Meiguang, of the Research Institute of Building Materials, made EDAX analyses of some of the Han Dynasty glass (Table 2). It can be seen that there were several different types of glass in the Guangxi tombs. Only one example of PbO-BaO glass was found, the blackish-green ear spool from the tomb in Zhaoping, Guangxi. Most of the glasses were K_2O-SiO_2 glasses. Whether or not this kind of glass was made in China is still not clear.[19] The shape and the color of the ear spool at Guixian are similar to those of ear spools excavated at Zhaoping, but the analysis shows that the Guixian object is not a PbO-BaO glass, but a K_2O-SiO_2 glass. We believe, as do other authors,[20, 21, 22] that K_2O-SiO_2 glasses were made in China.

We cannot rule out the possibility that some objects were imported. During the Qin Dynasty, Guangxi became a friendly line of communication between China and Southeast Asia. Hepu, Guangxi, was then an important harbor. In *Wei Shu. Xi Jong Zhuan,* glass made in Daqin Guao was described in detail. Daqin Guao produced ten kinds of glass: red, white, black, green, bluish-green, dark purple, bluish-green and white, crimson, and purple. The colorful glasses naturally were attractive to the ruler of the Han Dynasty, and Emperor Hanwu spared no expense to dispatch officials to foreign countries to buy them. Therefore, some experts believe that some of the glasses excavated at Guixin were imported.

In summary, we can make three points:

1. Among the glass excavated from the Han Dynasty tombs in Guangxi, only one example of a lead-barium object was found.

2. Judging from the shapes and patterns of the cups and ear spools, a comparatively large quantity of potash-silica and potash-lime-silica glass was made in China, but we cannot rule out the possibility that some of that glass was imported from foreign countries.

3. It is certain that ancient glass objects were imported into Guangxi, which is situated on the sea in southwest China, and that there were frequent contacts between foreign countries and Guangxi.

Notes

1. Fan Shimin and Zhou Baozhong, "A Study of Some of the Glass Objects Stored in the Museum," *Bulletin of the Museum of Chinese History,* no. 5, 1983, pp. 97-104.
2. An Jiayao, "Early Glass Vessels of China," Chapter 1, this volume.
3. Gan Fuxi, Huang Zhenfa, and Xiao Bingron, "The Origin of Ancient Chinese Glass," *Journal of the Chinese Silicate Society,* v. 6, nos. 1 and 2, 1978, pp. 99-104.
4. Han Shu, "Pearls, Ambers and Glass *Bi* Produced in Va Qin Yanbing Guao," *Zi Yu Zhuan.*
5. Tai Pin Yu Lann, "Five-color Glass Produced in Da Qin Guao," *Xun Zhong Ji,* v. 808.
6. Gan, {Note 3}.
7. Fan and Zhou, {Note 1}.
8. Fan and Zhou, {Note 1}.
9. Gan, {Note 3}.
10. Fan and Zhou, {Note 1}.
11. The Museum of Boxian, Anhwei Province, "Tombs of Caocao's Clan at Boxian," *Cultural Relics,* no. 8, 1978.
12. Gao Xian, "Bieguao Dong Ming Ji."
13. Gao, {Note 12}.
14. Bi Yuan, "San Pu Huang Tu."
15. Wu Jun, "Miscellanies in Xi Jing."
16. Tokyo National Museum, "Japan's Ancient Glass from the Viewpoint of Contacts Between the East and the West."
17. Tan and Sialichen, "Recent Archaeological Finds in Malaysia in 1964," *Bulletin of the Royal Asian Association, Malaysia Branch,* v. 28, no. 1, July 1965.
18. Fan and Zhou, {Note 1}.
19. Zhang Fukang, Cheng Zhuhai, and Zhang Zhigang, "An Investigation of Ancient Chinese '*Liuli*'," *Journal of the Chinese Silicate Society,* v. 11, no. 1, 1983, pp. 67-76.
20. Shi Meiguang, of the China Building Materials Academy, has told us that the glass ear spools excavated in Guixian are made of a potassium glass. Shi Meiguang, He Ouli, and Zhou Fuzheng, "Investigation of Some Chinese Potash Glasses Excavated from Tombs of the Han Dynasty," *Journal of the Chinese Silicate Society,* v. 14, no. 3, 1986, pp. 307-313.
21. Chen Dachang, "Yan Fan Lu."
22. Robert H. Brill, "Chemical Analyses of Some Early Chinese Glass," *Silicate Digest,* no. 7, 1984.

Table 1

Glass Objects Excavated in Guangxi

Date	*Location*	*Type*	*Brief Description*	*Source*
Early Western Han Dynasty	Gaochai, Hexian	30 beads	Round, deep, blue	*Cultural Material Collection*, no. 4, 1981
Early Eastern Han Dynasty	M1, Loaowan, Guixian	1 bead, 1 ear spool, broken glass	Bead: olive-shaped Ear spool: Green translucent	Guangxi Museum
Late Western Han Dynasty	M1, Yuchang, Guixian	1 bead	Blackish-green, round	Guangxi Museum
Late Western Han Dynasty	M3, Xinniuling, Guixian	1 bead	Green, oval, with edge	*Cultural Relics, Reference Materials*, no. 2, 1957
Late Western Han Dynasty	M1, Wangniuling, Hepu	725 beads	Round, blue, blackish-green	Guangxi Museum
Late Western Han Dynasty	M2, Wangniuling, Hepu	1 *bi*	Blackish-green, trellis design	Guangxi Museum
Western Han Dynasty	M1, Tangpai, Hepu	437 beads	Green, sky-blue, light bluish-green, blackish green	*Cultural Material Collection*, no. 4, 1981
Western Han Dynasty	M2, Tangpai, Hepu	2 beads		*Cultural Material Collection*, no. 4, 1981
Late Western Han Dynasty	M2B, Tangpai, Hepu	131 beads, 5 tubes	Bead: blue, green Tube: blue	*Cultural Material Collection*, no. 4, 1981
Late Western Han Dynasty	M3, Tangpai, Hepu	1 bowl, 1084 beads	Bead: sky-blue, lake-water blue, green	*Cultural Material Collection*, no. 4, 1981

Late Western Han Dynasty	Gaochai, Hexian	22 beads	Green, black, white, flat	*Cultural Material Collection*, no. 4, 1981
Eastern Han Dynasty	M2, Fengliuling, Guixian	2 cups	Green opaque (1) Transparent (1)	Guangxi Museum
Eastern Han Dynasty	M5, Fengliuling, Guixian	1 bead		Guangxi Museum
Eastern Han Dynasty	Guixian	1054 beads, 1 bowl	Bead: abacus, bluish-green, blackish-green Bowl: broken	*Archeology*, no. 1, 1957
Eastern Han Dynasty	M8, Nandao Village, Guixian	1 cup	Short stem, saucer, light bluish-green, transparent	Excavated in Guangxi
Eastern Han Dynasty	Railway station, Guixian	1 cup	Wide-mouthed, flat bottom, blackish-green opaque	Excavated in Guangxi
Eastern Han Dynasty	Guixian	1 cup	Wide-mouthed, green, transparent	*Journal of the Museum of Chinese History*, no. 5, 1983
Eastern Han Dynasty	Guixian	1 bead		*Cultural Relics Reference Materials*, no. 2, 1955
Eastern Han Dynasty	Wenjingling, Guixian	1 bead		*Cultural Relics Reference Materials*, no. 2, 1956
Eastern Han Dynasty	Bamen, Guixian	1 bead		*Kaogu*, no. 4, 1956
Eastern Han Dynasty	M7, Northern suburb, Guixian	1 ear spool	Blackish-green, waist drum-shaped	Guangxi Museum

Eastern Han Dynasty	M5, Northern suburb, Guixian	39 beads	Light bluish-green, green, sky-blue, brick-red, blackish-green, purple brown	Guangxi Museum
Eastern Han Dynasty	M5, Le, Zhaoping	14 beads, 2 ear spools	Bead: Sky-blue, green, Ear spool: Blackish-green	Guangxi Museum
Eastern Han Dynasty	M6, Le, Zhaoping	2 beads	Sky blue, lake-water blue	Guangxi Museum
Eastern Han Dynasty	M7, Le, Zhaoping	42 beads	Green, sky-blue, brick-red, white	Guangxi Museum
Eastern Han Dynasty	M8, Le, Zhaoping	15 beads	Blackish-green, sky-blue green	Guangxi Museum
Eastern Han Dynasty	M9, Le, Zhaoping	1 bead	Blackish-green	Guangxi Museum
Eastern Han Dynasty	M10, Le, Zhaoping	1 bead	Sky-blue, abacus	Guangxi Museum
Eastern Han Dynasty	M11, Le, Zhaoping	1 bead	Sky-blue, abacus	Guangxi Museum
Eastern Han Dynasty	M14, Feng, Zhaoping	1 bead	Blackish-green	Guangxi Museum
Eastern Han Dynasty	M5, Feng, Zhaoping	1 bead	Sky-blue	Guangxi Museum
Eastern Han Dynasty	M8, Feng, Zhaoping	11 beads	Green, sky-blue	Guangxi Museum
Eastern Han Dynasty	M14, Feng, Zhaoping	1 bead	Sky-blue	Guangxi Museum
Eastern Han Dynasty	M1, Shimaping, Xingan	100 beads	Blue, white, abacus	Guangxi Museum
Eastern Han Dynasty	M10, Shimaping, Xingan	Glass ornament		Guangxi Museum
Eastern Han Dynasty	Liljong, Liujiang	1 bead		Guangxi Museum
Eastern Han Dynasty	M2, Hetousan, Wuzhou	200 beads	Light green	*Cultural Relics Collection*, no. 4, 1981

Table 2

Chemical Analyses of Glass from Guangxi

No.	Object	Location/Date	Chemical Composition											
			SiO_2	Al_2O_3	Fe_2O_3	TiO_2	CaO	MgO	K_2O	Na_2O	MnO	Cl	PbO	BaO
1	Blue bead	Western Han Dynasty (Yandui, Hepu)	81.2	2.69	0.65		1.0	0.49	12.16	0.79	0.36			
2	Blue bead	Western Han Dynasty (Hepu)	78.22	2.56	1.28	0.14	1.45	0.27	13.81	0.41	1.85			
3	Cup	Western Han Dynasty (M2, Feng, Guixian)	77.50	4.03	0.77			0.69	15.38	1.53		0.1		
4	Green Cup	Western Han Dynasty (M2, Feng, Guixian)	74.66	6.24	0.27		0.76	0.58	15.66	0.56	1.52	0.75		
5	Green Bead	Eastern Han Dynasty (M2, Feng, Zhaoping)	83.9	2.39			1.18		11.03	1.38				
6	Blackish-green eardrop	Eastern Han Dynasty (M5, Le, Zhaoping)	35.04	1.99	0.345		2.67	2.24	0.27	5.1			22.3	8.16
7	Green Cup	Eastern Han Dynasty (Railway station, Guixian)	+++		+		+++		++					=

No. 7 was analyzed qualitatively by X-ray fluorescence in 1982 by Chang Riching and Qu Changzhi of the Physics Engineering Department of Ching Hua University. Nos. 1-6 were analyzed by EDAX in August, 1984 by Shi Meiguang of the China Building Materials Academy.

CHAPTER 19

SOME GLASS IN THE MUSEUM OF CHINESE HISTORY

FAN SHIMIN AND ZHOU BAOZHONG

Museum of Chinese History,
Beijing

GLASS objects of many periods have been unearthed in China, and they have captured the attention of historians, archaeologists, and natural scientists. As a result, research on early glass in China is progressing rapidly. Because historical records from the Western Han Dynasty onward refer to glass objects having been imported into China from middle and western Asia, as well as from the Eastern Mediterranean, two important questions arise. First, did glass have an independent origin in China, and second, what was the relationship between Chinese glass technology and glass technology of other parts of the world in ancient times?

In this paper we express our initial views on this topic and discuss some of the glass in the Museum of Chinese History.

The Western Zhou Dynasty (11th-8th c.)

In recent years many beads have been unearthed at Rujia Zhuang in Baoji, Shaanxi. These prompted some researchers to put forwrd their views as to the origin of Chinese glass objects, but their views have not led to a consensus of opinion in academic circles. However, glass objects in our museum's collection may confirm that the Western Zhou Dynasty was the embryonic period of glassmaking in China. Until now, the beads excavated in the Western Zhou tomb at Zhongzhou Road in Luoyang, Henan Province, are the earliest glass artifacts unearthed in China (Fig. 1). They were excavated in the 1950s by the Institute of Archaeology of the Chinese Academy of Sciences. The beads are light green in color, and are heavily weathered. Their shapes resemble irregular spheres about 0.5 cm in diameter with bores about 0.3 cm. Spectrographic analyses by the Institute of Applied Physics of the Chinese Academy of Sciences suggested that the beads were made of glass, but because the analyses were incomplete, we sent samples to Zhang Riqing and Qu Changzhu of the Department of Engineering Physics of Qinghua University in May, 1982. They carried out x-ray fluorescence analyses. Their qualitative analyses show that the beads

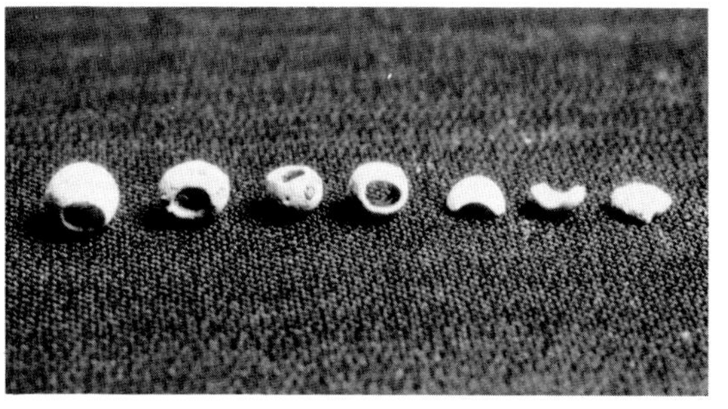

Figure 1. *Western Zhou beads from Luoyang.*

contain a major amount of silica and minor amounts of iron and copper. The remaining oxides are those of calcium and potassium.

The main question now is whether the beads in the tomb of the Western Zhou Dynasty at Luoyang were imported from western countries or were made in China.

It is generally believed that glasses made in western countries at that time were soda-lime-silica glasses in which sodium carbonate was used as a flux. When such glass objects became weathered, the sodium and some other elements were dissolved out by ground water, leaving thin weathered layers consisting primarily of silica. (The longer the weathering time, the more the weathered layers.) Are the beads of Zhongzhou Road this kind of material? The x-ray fluorescence analysis did not show any sodium, and we could not see weathering layers through the microscope. Therefore, we believe the beads are not western glasses which have become weathered. Perhaps they are, as some observers have decided, not really made of glass but rather of faience.

In the 1940s, the English scholar H. C. Beck and others studied the glass artifacts excavated in tombs of the Warring States Period in Luoyang. They found that the glasses were different from those made in western countries in that they contained much lead and barium oxides. After that, other researchers, such as Prof. Yuan Hanqing of Beijing Teachers University, carried out further studies. Prof. Yuan analyzed three glass beads from a tomb of the Warring States Period (475-221 B.C.) in Luoyang and found that the beads contained 25.03-74.61% lead oxide and 5.92-13.29% barium oxide. This research helps in distinguishing the glass of China of the Warring States Period from that of foreign countries of the same period. However, the Warring States Period was not the embryonic period of Chinese glass technology. So, the problem of glass production before the Warring States Period can not be appraised by looking only at the oxides of lead and barium. If the glass beads of the Western Zhou Dynasty unearthed at Zhongzhou Road are regarded as having been produced in China, it seems that non lead-barium glass was produced in China before the lead-barium glass artifacts brought into being during the Warring States Period. Then the new and important question of locating the area which produced the earlier glass is raised. All these pending questions need further analysis and study.

Glass of the Warring States Period

According to records in the book *Strategies of the Warring States - Strategies of Chu,* Qin sent a messenger to Huai (King of Chu) and subsequently King Huai "...sent one hundred messenger carts to present rhinoceros horn with colored veins and a luminous *bi* disk from Qin." There was a sentence describing, "the luminous colored glaze" in the book *The Compose* {?} *of Feather Hunting* by Yang Xong. From this we know that the so-called luminous *bi* was actually made of glass. Many glass items of the Warring States Period have been unearthed in Henan, Hebei, Shandung, Shaanxi, Hunan, and Hubei. The kinds of articles we have seen include beads, *bi*, rings, and other forms. Among these, the glass beads are most numerous and are well made. *Bi*, rings, and other items engraved with clouds, dragons, and grain are in traditional Chinese style. But the designs of the glass beads are very similar to those of glass beads made in the West. That was why some Chinese researchers thought foreign glass objects had been imported into China in ancient times. Such speculation is bolstered by the fact that in the 4th century B.C., Alexander the Great led his army to the East and established a mighty empire which covered parts of the three continents: Europe, Asia, and Africa; his Empire's sphere of influence reached as far east as the Indus Valley. Glass artifacts as well as objects inlaid with glass were spread from western countries to many parts of West Asia and Central Asia by Alexander's conquests. The question as to whether the color and design of the glass beads of the Warring States Period were influenced by the arrival of the same kind of beads made in western countries is worth further inquiry, but the chemical compositions of the glass objects made in China indicate that this Chinese glass is distinctly different from those made in the West.

The following are X-ray fluorescence analyses of the four glass beads in our Museum:

1. A hexagonal, tubular-shaped glass bead (unearthed in 1951-1952, in tomb 1 at Guwei Village in Hui Xian County, Henan Province). Transparent blue; length 2.2 cm, 1.3 cm in the middle, 0.9 cm at both ends. Contains high percentages of lead and barium oxides (Fig. 2).
2. A glass bead (same source as above). Green with blue dots and white circles; diameter 1.2 cm, 1.3 cm high. Contains high percentages of lead and barium oxides.
3. A *bibai*-shaped silver belt hook inlaid with gold, jade, and glass beads (same source as above). Glass beads inlaid in the middle of three jade rings on back. Diameter 1.2 cm. Contains high percentages of lead and barium oxides. (Fig. 3).
4. An inlaid glass bead (allegedly unearthed in a Warring States Period tomb in Changsha, Hunan Province). Tube-shaped. White body inlaid with different-sized yellow, blue, and black glass beads. Diameter 2.1 cm; diameter of bore 0.9 cm, height 1.8 cm. Contains high percentages of lead and barium oxides.

Figure 2. *Hexagonal tubular bead from Hui Xian County, Henan (on right).*

Figure 3. Bibai-*shaped belt hook with eye beads mounted in jade disks, from Hui Xian County, Henan.*

The high lead and barium oxide contents in these objects, and the data of Prof. Yuan described above, show that the materials used for glassmaking in China were different from those used in the West during the Warring States Period. In China, lead was discovered very early. During the Shang and Zhou Dynasties, not only was lead extracted from ores to make lead articles, but lead was also used for making metal alloys. As for the barium, it is found in pottery made in the Shang and Zhou Dynasties. [Editor's Note: No reference is cited for this information, but it is clearly of interest and perhaps of great importance.} Therefore, the use of lead and barium as a flux for making glass in China during the Warring States Period was not accidental.

Glass of the Western and Eastern Han Dynasties

During the two Han Periods (206 B.C.-220 A.D.), glass manufacturing developed rapidly. From archaeological finds, we know that glass objects not only increased in number and variety but also in their transparency. The book *On Measures - Rate Prosperities* {?} by Wang Chang says the Taoist priests "...smelt five kinds of stone to make flint. After polishing, when the flint was faced towards the sun, fire might be caused." Recently, two pieces of highly transparent lens glass were unearthed in a tomb in Caocao in Boxian County, Anhui Province. Researchers were surprised that such high quality glass objects had been produced so early. In addition, there are the Han Dynasty glass objects in the National Museum of Japan in Tokyo: one is a belt hook, the other a hexagonal glass "stick." (Japanese scholars call it *wo*.) The transparency of these might correspond to the description in Wang Chang's book.

During the two Han Dynasties, cultural exchange between China and the West grew more frequent. While elegant silk articles and high quality iron and steel products of China were exported westward in great quantities, glass from western countries (along with other goods) was imported into China over both land and sea trade routes. These glass objects were treated as treasures in China. According to *Documents of Han Dynasty, Geography* and *On Salt and Iron,* at that time glass objects were not only imported by foreign trade caravans, but China also sent messengers taking gold and silk abroad with them to buy glass. In the past few decades, archaeologists have discovered many glass objects along thc southeast coast. Thousands of glass beads were unearthed in the Han Dynasty tombs of Guangzhou. Among the imported glass articles, the blue glass bowl and the green glass cup in our Museum's collection attract special attention from our visitors.

Figure 4. *Blue bowl from Hengzhigang, Guangzhou. (Rim diameter 10.6 cm)*

Figure 5. *Green cup from Gui Xian County, Guangxi. (Rim diameter 5.8 cm)*

The blue glass bowl (Fig. 4) was unearthed in a wood coffin in tomb No. 1, dated to the Western Han Dynasty, at Hengzhigang in Guangzhou, in 1954. The green glass cup was unearthed in Gui Xian County, Guangxi Province, in 1957. Guangzhou and Gui Xian County have been important seaports and river harbors in China ever since the Qin and Han Dynasties.

The blue glass bowl has straight edges and a circular bottom with a thread beneath the rim. The inside surface is smooth and the outside rough. It is translucent. The bottom diameter is 4 cm and the diameter of the rim 10.6 cm.

The green glass cup has a wide opening and a folded rim; it has a circular bottom. There are two cut grooves near the waist. The whole body is smooth and transparent. It is 3.2 cm high and 5.8 cm at its mouth (Fig. 5). The chemical analysis shows that it does not contain lead, but is a potash-lime-silica glass. {Editor's Note: The text is not completely clear on this point.}

In 1957, two more glass objects were unearthed in Gui Xian County: one is a green glass cup, exactly the same shape and design as the one in our Museum. The other is a goblet with a green glass saucer. Their shapes and designs are quite different from traditional Chinese glass objects. Instead, they have characteristics of glass made in the Near East. For this reason, the green glass cup in our Museum, and these two glass cups, might all have been imported from the Near East over the long sea route. As for the source of the blue glass bowl, it is difficult for us to judge at the present time. {Editor's Note: Having handled both these objects, one of the editors (R.H.B.) believes that the green cup (actually, it is aqua in color) is just as likely to have been made in China, while the source of the blue bowl is still problematical. The qualitative analyses suggest that the green cup is a potash glass, reinforcing a supposition of a Chinese origin. The blue bowl contains at least some lead and barium. Sodium is not reported for either glass but it was probably not detectable by the instrument used for the analyses. Full quantitative analyses would probably solve the questions of the origins of both glasses unambiguously.}

Glass of the Wei, Jin, Southern and Northern Dynasties

In the 1940s, the 1st-3rd century ruins of Begram were excavated in Afghanistan. Large quantities of Roman blown glass were unearthed. During the Wei and Jin Dynasties, such high quality blown glass also came eastward into China across the Kunlun Mountains from the Kushan Empire. During the Western Jin Dynasty, Pan Ni in his *Poem of the Glass Bowl* wrote, "...the glass treasures, of which the glass bowl was the most magnificent, crossed the dangerous sand hills and climbed over the high Congling Mountains." The glass cup unearthed in the Wang tomb of the Eastern Jin Dynasty at Xiangshan in Nanjing in the 1970's, and the few high quality blown glass objects unearthed in the tomb of Feng Sufu of the Northern Yan Dynasty at Beipiao in Liaoning Province, might have come to China from the Roman Empire.

Figure 6. *Cup with network decoration from Jing Xian County, Hebei.*

Two glass objects in our Museum were unearthed in the Feng tombs in Jing Xian County, Hebei Province, in 1948. One of them was a glass cup with applied network decoration (Fig. 6). Its shape was new and strange, and it has the appearance of glass made in the West. It contains no lead or barium, and it might have been imported from the Mediterranean or Persia.

During this period, references to learning from Western glass technology appeared in Chinese literature. The book *Baspuzi (Yellow and White)* written by Ge Hong says, "...the water demon bowls made in foreign countries were actually made of five different clays. {Components?}. Now such things are made using their methods in Guangdong Province; but people do not believe such things are man-made. They believe the water demons {?} are made of a natural material. It is a kind of jade, just as there is native gold in the world. How can people make natural matter?" The "water demon" of "the natural matter" in Ge Hong's book must have meant crystal from quartz mines, and the "water demon bowl" meant the glass bowls made of five materials. What were the five kinds of material? Ge Hong did not name them. The materials used in making glass in China were still the traditional materials, but the method of forming might have been influenced by the technology of the western countries.

The other object is a shallow green glass bowl, 4.5 cm in height, and 11.2 cm at its rim with only a little lead in it. The shape is the same as Chinese traditional utensils. It might be a glass made in China following an example of foreign methods (Fig. 7).

Figure 7. *Green bowl from Jing Xian County, Hebei. (Rim diameter 11.2 cm)*

Sui Dynasty

During the unification of the south and north of China, glass manufacture stagnated, but it was to resume and develop again. The book *Documents of the Sui Dynasty, He Chou Biography* says, "At the beginning of the first Emperor's reign ... glass (*liuli*) manufacture had stopped for a long time. Craftsmen did not dare to disobey the emperor's order. He Chou, a craftsman, adopted the method of making green china {?} to make glass articles, and they were exactly like the real glass articles." What the items "exactly like the real thing" were, nobody knows.

The eight glass objects unearthed in the tomb of Li Jingxun in Xian in 1957, and now in our Museum, supplied us with material to study glass manufacturing during the Sui Dynasty (581-618). They were vases, pots, cups, a tube, and an egg-shaped object. Four were analyzed qualitatively by X-ray fluorescence. These included a glass cup with folded rim and flat bottom, a small pot with a cover, and another cup. The pot is thin and transparent and very well made. Although buried a long time in the earth, it is still smooth and shiny. It has a high lead content, and it may be the first blown glass article made in China.

The appearance of blown glass was an important turning point in the history of glassmaking. It made mass production possible and the glass products less expensive. It is believed that glassblowing was invented in the Middle East, perhaps in Syria, about 50-25 B.C. In China, however, glassblowing did not develop as early as in the West. Glass made after the Sui and Tang Dynasties fell far behind Western glass both in quantity and quality. The reasons for so late a development of glassblowing warrants further study.

Glass objects had been luxuries enjoyed by only a few nobles and rulers in ancient China. The limited scope of use and the meager demand might have been the social reasons for its slow development. In ancient China, glasswares were far inferior to porcelain in quality. Glass containers broke easily, for the physical properties of ancient Chinese lead and barium glass made it unsuitable for use at high temperatures and prone to breakage caused by sudden changes in temperature. Thus, Chinese glass objects were not as practical as those made in Western countries. Western glass had the same physical qualities as Chinese china and silver articles where boiling water would not cause them to break. Because of these practical shortcomings, the manufacture of glass in ancient China, therefore, did not reach the same advanced level as in the West.

To sum up, it is our opinion that:
1. During the Warring States Period, the typical Chinese lead-barium-silica glass technology was already in use.
2. Since the Western Han Dynasty, glass objects made in western countries were imported into China. Chinese glass manufacture was influenced by Western advanced production technology.
3. The beginning of glass manufacture in China was later than its beginning in the West. However, the use of lead and barium as fluxes was a forerunner of the modern manufacture of lead-glass.

Acknowledgements. The chemical analyses mentioned in this article were provided by Zhang Riging and Ou Changzhi, of the Department of Engineering Physics, Qinghua University, by An Jiayao of the Institute of Archaeology, and by Han Fulin of the Museum of Chinese History. We express our sincere thanks to them.

Table 1
Chemical Analyses of Some Chinese Glasses*

Number	*Type of Object*	*Source and Date*	*Si*	*K*	*Ca*	*Pb*	*Ba*	*Fe*	*Cu*
1	Green bead	Luoyang; 11th-8th c. B.C.	+++	++	++			+	+
2	Inlaid bead	Guwei village, Huixian County, Henan; 5th-3rd c., B.C.	++	++	++	+++	+++	+	+
3	Hexagonal bead	same	++	++	++	+++	+++	+	+
4	Bibei	same	++		++	+++	+++	+	+
5	Inlaid bead	Changsha; poss. 5th-3rd c. B.C.	++	++	++	+++	+++	+	+
6	Blue bowl	Guangzhou; 2nd-1st c. B.C.	++	++	++	+	+	+	+
7	Green cup	Guixian County, Guangzhou; 1st-2nd c. A.D.	+++	++	+++		+	+	
8	"Network" cup	Jingzian County, Hebei; 5th-6th c. A.D.	+++	++	+++			+	+
9	Green bowl	same	+++	++	+++	+		+	+
10	Flat-bottomed vase	Xian; 6th-7th c.	+++	++	+++		+	+	
11	Cup	same	+++	++	+++			+	
12	Cup	same	+++	++	+++		+	+	
13	Pot with cover	same	+++	++	++	+++			

*All x-ray fluorescence analyses, by Zhang Riqing and Qu Changzhu, 1982.

CONTRIBUTORS

Because many of the authors and others connected with this book may never meet in person — and because most will be unknown to the readers — we are presenting here photographs of as many contributors as we could obtain. Also included are some of the people who were pioneers in the scientific investigation of early glasses. Except as noted below, the photographs were provided by the individuals themselves or taken by the editors. A few of the authors declined our invitation to submit photographs.

Photo credits: Horace C. Beck, Mrs. Flora Westlake; W. E. S. Turner, The Society of Glass Technology; Earle R. Caley, Dr. Aaron Ihde; M. A. Besborodov, Prof. Kazuo Yamasaki.

Participants in TC-17 Symposium, Beijing, September 7, 1984. Li Jiazhi, Liu Tongzhao, Horst Scholze (President, ICG), Yi Jialiang, John Wosinski, Shi Meiguang, Kazuo Yamasaki, Robert Brill, Gao Zhixi, Cheng Zhuhai, Huang Zhenfa, Zhou Changyuan (kneeling).

Gan Fuxi (Chairman, Symposium Program Committee).

Kazuo Yamasaki and William Prindle (Vice President, ICG).

He Ouli

Gong Fangtian

Shi Meiguang

Fan Shimin

Xiong Zhuanxin

Kazuo Yamasaki

W.E.S. Turner (about 1930).

An Jiayao, Xia Nai, and Lu Zhaoyin (Beijing, April 17, 1982).

Horace C. Beck (about 1913).

Earle R. Caley (about 1957).

Abdugani A. Abdurazakov (Tashkent, July 9, 1989).

Michael A. Besborodov (Leningrad, July 4, 1973).

Liu Tongzhao

Philip M. Fenn

Huang Qishan

Masayo Murozumi

Fan Jianchun

John F. Wosinski

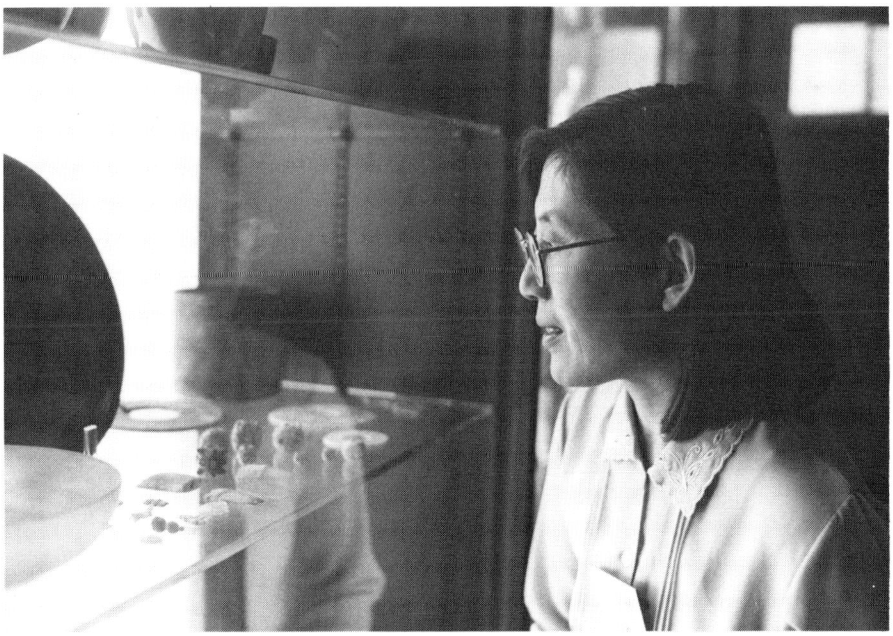

An Jiayao

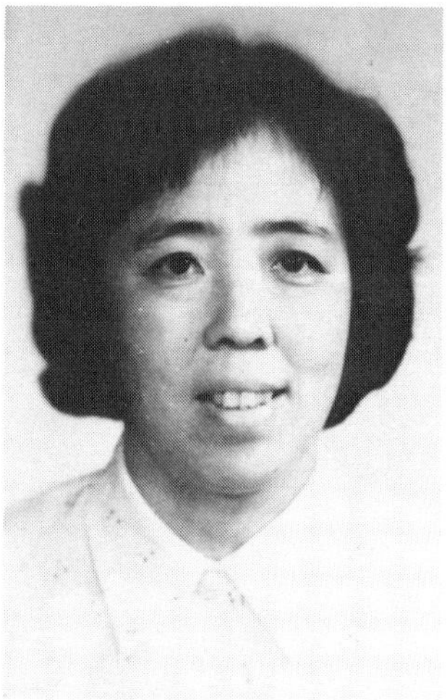

Zhou Fuzheng

Wu Zhongdao

James C. Y. Watt

Jia Shaoping and Lu Shuolin.

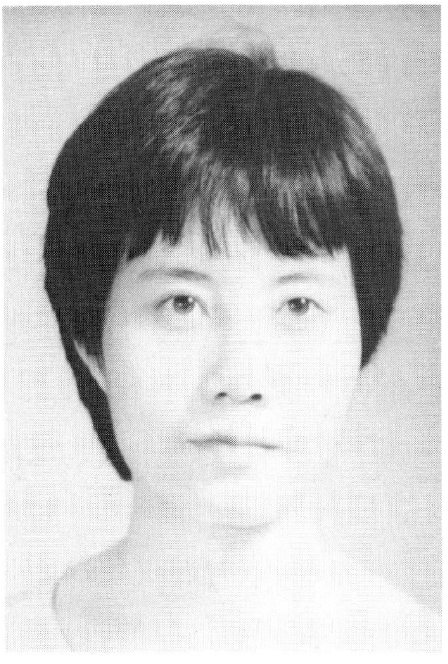

Wang Shixiong

Doris Dohrenwend

Stephen S. C. Tong

Elizabeth R. Brill and Pierre Gilard (TC-17).

Yang Boda

Robert D. Vocke, Jr.

Zhang Fukang

Margery Chen

I. Lynus Barnes

Pamela England

In-Sook Lee and Maxine Smith.

Ruth C. Sprague

John H. Martin

Mary L. Townsley

Richard W. Price